From corresponding states in water – CO_2 – $F(CF_2)_iC_2H_4E_j$ microemulsions towards foamable

nanostructured CO_2-in-polyol systems

From corresponding states in water $-$ CO_2 $-$ $F(CF_2)_iC_2H_4E_j$ microemulsions towards foamable nanostructured CO_2-in-polyol systems

Von der Fakultät Chemie der Universität Stuttgart
zur Erlangung der Würde eines
Doktors der Naturwissenschaften (Dr. rer. nat.)
genehmigte Abhandlung

Vorgelegt von

Stefan Lülsdorf

aus Troisdorf

Hauptberichter: Apl. Prof. Dr. Thomas Sottmann
Mitberichter: Prof. Dr. Reinhard Strey
Prüfungsvorsitzender: Prof. Dr. Michael R. Buchmeiser

Tag der mündlichen Prüfung: 20.07.2018

Institut für Physikalische Chemie der Universität Stuttgart
2018

Bibliografische Information der Deutschen Nationalbibliothek

Die Deutsche Nationalbibliothek verzeichnet diese Publikation in der
Deutschen Nationalbibliografie; detaillierte bibliographische Daten sind im Internet
über http://dnb.d-nb.de abrufbar.
1. Aufl. - Göttingen: Cuvillier, 2019
    Zugl.: Stuttgart, Univ., Diss., 2018

D93 (Diss. Universität Stuttgart)

© CUVILLIER VERLAG, Göttingen 2019
    Nonnenstieg 8, 37075 Göttingen
    Telefon: 0551-54724-0
    Telefax: 0551-54724-21
    www.cuvillier.de

    ISBN 978-3-7369-7031-1
    eISBN 978-3-7369-6031-2

Abstract

Today there is an increased focus on a considerable reduction of the total energy consumption in the world. As the heating of buildings constitutes a large part of this consumption, the development of highly efficient thermal insulation materials is highly demanded. The most promising class are polymeric nanofoams, which might be producible by using the "Principle Of Supercritical Microemulsion Expansion" (POSME). Within this approach, microemulsions with a high number density of CO_2 nano-pools in a polymerizable material (e.g. polyol) are used as templates. Adjusting supercritical conditions, the nucleation of gaseous CO_2 during the expansion is avoided, so that each CO_2-swollen micelle should ideally grow to a nanopore. However, due to aging phenomena, so far only polyurethane (PU) foams with microcellular pores could be produced applying this approach. In order to improve the properties of the foamable microemulsion template, microemulsions of the type water – CO_2 – non-ionic fluorinated surfactant (Capstone® FS-3100) were studied in the first part of this thesis as a model system. Small angle neutron scattering (SANS) measurements for the first time revealed the phase inversion from CO_2-swollen micelles in water via bicontinuous structures to water-swollen micelles in CO_2. Using a scaling description developed for microemulsions of the type water – alkane – C_iE_j, the temperature dependence of the length scales found in classical and CO_2-microemulsions fall on top of each other. Moreover, the partial replacement of CO_2 by cyclohexane was found to increase the efficiency of the fluorinated surfactant to form a bicontinuous microemulsion. This efficiency boosting could be correlated with the increase of the bending rigidities $\kappa_{0,SANS}$ and κ_{NSE} determined by SANS and neutron spin echo (NSE). In the second part of this thesis the gained knowledge was successfully used to formulate CO_2-in-polyol microemulsions appropriate for the production of nanocellular PU foams. Using SANS and partly deuterated polyol, not only the formation of micelles in the binary mixture of polyol and non-ionic siloxane surfactant, but also their swelling with CO_2 could be proven. The change of the microstructure upon fast expansion and compression was studied combining time-resolved SANS measurements and periodic pressure jumps. Interestingly, the pressure induced structural changes follow the pressure profile instantaneously, i.e. faster than the time resolution of 50 ms.

Kurzzusammenfassung

Eine erhebliche Reduzierung des weltweiten Gesamtenergieverbrauchs, z.B. durch hocheffiziente Wärmeisolationsmaterialien, steht heutzutage immer mehr im Fokus. Polymere Nanoschäume, welche mittels des sogenannten POSME-Prinzips ("Principle Of Supercritical Microemulsion Expansion") hergestellt werden könnten, stellen das vielversprechendste Material dar. Beim POSME-Verfahren werden Mikroemulsionen mit einer hohen Anzahldichte an CO_2-*nano-pools* in einem polymerisierbaren Material (z.B. Polyol) als Templat verwendet. Durch die Verwendung von überkritischen Bedingungen wird die Nukleation von gasförmigen CO_2 während der Expansion vermieden, sodass idealerweise jede CO_2-geschwollene Mizelle zu einer Nanopore anwächst. Aufgrund von Alterungsphänomenen konnten bislang jedoch nur mikrozellulare Polyurethan (PU)-Schäume hergestellt werden. Um die Eigenschaften von verschäumbaren Mikroemulsionstemplaten zu verbessern, wurden in dem ersten Teil dieser Arbeit Mikroemulsionen des Typs Wasser – CO_2 – nicht-ionisches fluoriertes Tensid (Capstone® FS-3100) als Modellsysteme untersucht. Kleinwinkelneutronenstreumessungen (SANS) zeigten erstmals die Phaseninversion von CO_2-geschwollenen Mizellen in Wasser über bikontinuierliche Strukturen hin zu Wasser-geschwollenen Mizellen in CO_2. Mittels der für Mikroemulsionen des Typs Wasser – Alkan – C_iE_j entwickelten korrespondierenden Zustandsbeschreibung fallen die Temperaturabhängigkeiten der Längenskalen für klassische und CO_2-Mikroemulsionen auf eine gemeinsame Kurve. Darüber hinaus konnte gezeigt werden, dass der teilweise Austausch von CO_2 durch Cyclohexan die Effizienz des verwendeten fluorierten Tensids für die Ausbildung einer bikontinuierlichen Mikroemulsion steigert, was mit einer Zunahme der Biegesteifigkeiten $\kappa_{0,SANS}$ und κ_{NSE} korreliert. Diese wurden mit Hilfe von SANS- und Neutronen-Spin-Echo (NSE)-Messungen bestimmt. Im zweiten Teil dieser Arbeit wurde das gewonnene Wissen genutzt, um erfolgreich CO_2-in-Polyol Mikroemulsionen zu formulieren, welche sich für die Produktion von nanozellulären PU-Schäumen eignen. Anhand der Verwendung von SANS und teilweise deuteriertem Polyol konnte nicht nur die Bildung von Mizellen innerhalb binärer Mischungen aus Polyol und nicht-ionischen Siloxan-Tensiden, sondern auch das Anschwellen dieser Mizellen mit CO_2 nachgewiesen werden. Die Änderung der Mikrostruktur auf eine rasche Expansion und Kompression wurde durch die Kombination von zeitaufgelösten SANS-Messungen mit periodisch stattfindenden Drucksprüngen untersucht. Interessanterweise folgten die druckinduzierten strukturellen Änderungen dem vorliegenden Druckprofil unmittelbar, d.h. innerhalb der Zeitauflösung von 50 ms.

Acknowledgement

Die vorliegende Arbeit wurde im Zeitraum von Januar 2015 bis Juni 2018 am Institut für Physikalische Chemie der Universität zu Köln (Januar-Juni 2015) und anschließend am Institut für Physikalische Chemie der Universität Stuttgart (Juli 2015 bis Juni 2018) unter wissenschaftlicher Anleitung von Herrn *Apl. Prof. Dr. Thomas Sottmann* angefertigt. In diesem Zusammenhang möchte ich ihm dafür danken, dass er es mir ermöglicht hat, das spannende und komplexe Thema der Polyol-reichen bzw. wässrigen CO_2-Mikroemulsionen zu bearbeiten. Seine stete Diskussionsbereitschaft und seine Ideen haben maßgeblich zum Erfolg dieser Arbeit beigetragen. Bedanken möchte ich mich auch für die Möglichkeiten, an der Summer School in Bombannes, verschiedenen (inter-)nationalen Konferenzen und zahlreichen Messreisen ans ILL teilzunehmen. Vielen Dank für die vielen Denkanstöße und die lustige und unvergessliche Zeit!

Bei Herrn *Prof. Dr. Reinhard Strey* möchte ich mich für die Übernahme des Zweitgutachtens meiner Dissertation, die angeregten Diskussionen und die Möglichkeit an dem „nanoPUR"-Projekt mitzuarbeiten bedanken.

Meinem GRAD*US*-Mentor Herrn *Prof. Dr. Michael R. Buchmeiser* danke ich für die Betreuung im Rahmen von GRAD*US* und für die Übernahme des Prüfungsvorsitzes.

Den Local Contacts Herrn *Dr. Peter Lindner* und Herrn *Dr. Ralf Schweins* vom D11, Herrn *Dr. Lionel Porcar* vom D22 und Herrn *Dr. Ingo Hoffmann* vom IN15, am Institut Laue-Langevin in Grenoble (Frankreich), möchte ich für die Unterstützung bei den diversen Messzeiten und zahlreichen Diskussionen herzlich danken. Insbesondere Herrn *Dr. Ingo Hoffmann* danke ich für die tatkräftige Unterstützung während der beiden NSE-Messzeiten, sowie für die Hilfe bei der Auswertung der NSE-Daten. Ferner gilt mein Dank den Technikern Herrn *David Bowyer* (D11), Herrn *Mark Jacques* (D22) und Herrn *Claude Gomez* (IN15) für ihre Unterstützung. Zudem danke ich dem Institut Laue-Langevin für die Möglichkeit, Streuexperimente durchführen zu können und für die finanzielle Unterstützung.

Bei Herrn *Dr. Christian Hahn*, sowie Herrn *Dr. Paul Heinz* und der Covestro Deutschland AG (bzw. der früheren Bayer MaterialScience AG) möchte ich mich für die produktiven Treffen und Telefonkonferenzen im Rahmen des PuNaMi- bzw. „nanoPUR"-Projekts bedanken. Zudem gilt mein Dank Herrn *Dr. Hubert Kuhn* von CAM-D Technologies GmbH für das zur Verfügung stellen der MFD-Simulationen des Polyol-Systems. Das PuNaMi-Projekt und damit auch diese Arbeit, wurden durch das Bundesministerium für Wirtschaft und Energie gefördert (03ET1409C), wofür ich mich herzlich bedanken möchte.

Den Firmen DuPont, Lehmann&Voss&Co. KG, Evonik Industries AG und insbesondere der Covestro Deutschland AG/Bayer MaterialScience AG danke ich für die Bereitstellung der meisten in dieser Arbeit verwendeten Chemikalien.

Für die sehr produktive und häufig stressige Zeit während den SANS- und NSE-Messreisen ans Institut Laue-Langevin in Grenoble (Frankreich) möchte ich mich bei allen Mitgliedern der verschiedenen Messzeiten bedanken. Ohne die Unterstützung von Frau *Kristina Schneider*, Frau *Diana Zauser*, Frau *Shih-Yu Tseng*, Frau *Dr. Yvonne Reimann*, Frau *Dr. Lena Grassberger*, Herrn *Oliver Wrede*, Frau *Dr. Yvonne Hertle*, Herrn *Prof. Dr. Thomas Hellweg*, Frau *Sonja Dieterich*, Herrn *Prof. Dr. Frank Gießelmann* und insbesondere Herrn *Apl. Prof. Dr. Thomas Sottmann* gäbe es keine SANS- und NSE-Daten. Ein besonderer Dank gilt darüber hinaus Herrn *Harun Bilgili*, der mich mit einer „kleinen" Nachtfahrt vor einigen Stunden Stillstand bewahrt hat. Auch wenn die Messzeiten etwas zu häufig dazu führten, dass mitten in der Nacht die Membran der Zelle und/oder der Balg ausgetauscht werden musste, werde ich die Tag- und Nachtschichten mit euch sehr vermissen.

Bei der Elektronikwerkstatt des Instituts für Physikalische Chemie der Universität Stuttgart unter Leitung von Herrn *Boris Tschertsche* und der mechanischen Werkstatt des Instituts für Physikalische Chemie der Universität Stuttgart unter Leitung von Herrn *Thomas Weigand* möchte ich mich für die unzähligen Stunden bedanken, die unter anderem in die SHP-SANS Zelle und unsere Druckzellen eingeflossen sind und somit die Experimente erst ermöglicht haben. Ein großer Dank gilt dabei im speziellen Herrn *Boris Tschertsche* und Herrn *Daniel Relovsky*. Zudem möchte ich mich für die Fertigung der Druckzellen und für die Hilfe bei technischen Problemen bei der Werkstatt des Instituts für Physikalische Chemie der Universität zu Köln unter der Leitung von Herrn *Herbert Metzner* bzw. Herrn *Viktor Klippert* bedanken. Bedanken möchte ich mich auch bei Herrn *Thomas Michaelis* für die Einweisungen in den Aufbau der Druckzellen und deren (De-)Montage.

Für die kritische und aufmerksame Durchsicht dieser Arbeit möchte ich Herrn *Dr. Christian Hahn*, Frau *Alexandra Mengs*, Frau *Kristina Schneider*, Frau *Shih-Yu Tseng*, Herrn *Yaseen Qawasmi* und Herrn *Apl. Prof. Dr. Thomas Sottmann* ganz herzlich danken.

Bedanken möchte ich mich auch bei Frau *Katja Schotka*, Frau *Sina Wurtz* und Herrn *Manuel Wahl* für ihre Messungen und Arbeiten im Rahmen ihrer Praktika, welche zum Teil in meine Arbeit mit eingeflossen sind.

Für die kurzfristige Messung der ESI-Spektren möchte ich Herrn *Dr. Markus Kramer* und Herrn *Dipl.-Ing. Joachim Trinkner* aus dem Institut für Organische Chemie der Universität Stuttgart danken.

Meinen Bürokollegen/innen Herrn *Dr. Sébastian Andrieux*, Frau *Shih-Yu Tseng* und Herrn *Yaseen Qawasmi*, sowie den gern gesehenen Gästen Herrn *Dr. Jan Thater* und Frau *Kristina Schneider* möchte ich für eine sehr produktive Arbeitsatmosphäre, aber auch für eine unterhaltsame Zeit danken. Dazu beigetragen haben insbesondere auch die vielen Laborkollegen, insbesondere Herrn *Harun Bilgili* und Herrn *Julian Fischer*. Darüber hinaus möchte ich dem gesamten Arbeitskreis von Herrn *Prof. Dr. Reinhard Strey* in Köln, sowie der gesamten Arbeitsgruppen von Herrn *Apl. Prof. Dr. Thomas Sottmann* für die stets sehr gute Arbeitsatmosphäre danken. Hierfür möchte ich mich auch bei Frau *Prof. Dr. Cosima Stubenrauch* und ihrer gesamten Arbeitsgruppe bedanken.

Ein spezieller Dank gilt meinen Trainingspartnern/innen vom *TSV Schmiden*, die einen willkommenen Ausgleich zu der manchmal recht stressigen Zeit während der Dissertation geschaffen haben.

Insbesondere möchte ich mich auch bei meinen Eltern *Beate* und *Heinz Walter Lülsdorf* für die Unterstützung und das Vertrauen in mich bedanken.

Publications

The results presented in this thesis have in part been published (or will be published in the near future). In particular, excerpts and figures throughout this Ph.D. thesis where taken from the following sources:

Chapter 3:

- S. Lülsdorf, R. Schweins and T. Sottmann, manuscript prepared for submission.

Chapter 4:

- S. Lülsdorf, R. Schweins, S. Vogt, H. Kuhn, C.J. Hahn and T. Sottmann, manuscript prepared for submission.

In addition, the author has published the following work during his Ph.D., which is not included in this thesis:

- N. Grimaldi, P. E. Rojas, S. Stehle, A. Cordoba, R. Schweins, S. Sala, S. Lülsdorf, D. Piña, J. Veciana, J. Faraudo, A. Triolo, A. S. Bräuer and N. Ventosa, *ACS Nano* **2017**.

- O. Wrede, Y. Reimann, S. Lülsdorf, D. Emmerich, K. Schneider, A.J. Schmid, D. Zauser, Y. Hertle, A. Beyer, R. Schweins, A. Gölzhäuser, T. Hellweg and T. Sottmann, submitted to *Chemistry of Materials*.

Contents

1 Introduction

Liquid and solid foams have a wide area of applications in everyday life. While liquid foams are for example used as fire-fighting foams, the latter are used as building material, foam rubber, for packaging or thermal insulation [1]. For thermal insulation materials, typically expanded or extruded polystyrene (PS) or polyurethane (PU) foams are used [2]. In general, PU foams are obtained by polyaddition of polyol and polyisocyanate, each exhibiting at least two functionalities, in the presence of a chemical and/or physical blowing agent [3]. In recent years, nanoporous materials with pore sizes in the nanometer range have gained increasing interest. Compared to non-porous materials and foams with pore sizes in the micrometer range, these nanoporous materials exhibit improved and new properties, e.g. optical transparency, mechanical stability and low thermal conductivity [4–7]. With respect to the envisaged production of high performance insulation materials, PU is particularly suitable, as it exhibits a large potential for improvements.

The insulating capacity of porous materials is quantified by the inverse of the thermal conductivity λ_{therm}, which comprises contributions of the entrapped gas λ_g, the solid material λ_m and the thermal radiation λ_{IR} [8]. State of the art polyurethane foams exhibit pore size of around 100 µm and a thermal conductivity of $\lambda_{therm} \approx 0.030$ W/mK [3, 9]. As the gaseous thermal conductivity makes the major contribution [9], the most promising approach to lower the thermal conductivity is a reduction of λ_g.

One possibility to reduce λ_g is the use of gases with a low gas phase thermal conductivity $\lambda_{g,0}$. In the case of PU foams, ozone-depleting chlorofluorocarbons (CFCs) with $\lambda_{g,0} \approx 0.008$ W/mK were used as blowing agent from the late 1950s until the mid-1980s [10, 11]. With respect to the environmental protection, nowadays PU is mainly foamed using CO_2 ($\lambda_{g,0} \approx 0.016$ W/mK) or pentane ($\lambda_{g,0} \approx 0.011$-$0.016$ W/mK) [2, 10].

Another method to reduce λ_g is the *Knudsen* effect (see Figure 1.1) [12–14]. If the pore size δ_p is comparable with or smaller than the mean free path length of an enclosed gas (~ 100 nm at atmospheric pressure), the probability for a gas molecule to collide with another gas molecule is lower than for a collision with the pore walls. This means, that the gaseous thermal conductivity approaches zero. Beside a reduction of the pore size, the *Knudsen* effect can also be achieved by reducing the mean free path length via a reduction of the gas pressure p_g (e.g. down to 1 mbar)

[15]. Figure 1.1 shows the gaseous thermal conductivity λ_g as a function of the gas pressure p_g and characteristic size of pores δ_p.

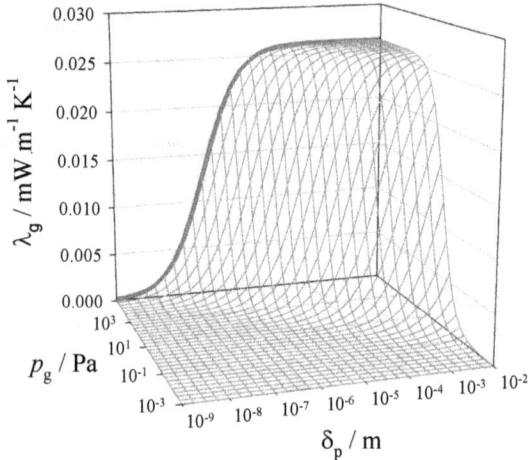

Figure 1.1: Calculated values for the gaseous thermal conductivity λ_g as a function of the gas pressure p_g and characteristic size of pores δ_p. λ_g was calculated via the *Knudsen* number with: $\lambda_{g,0} = 0.0179$ W/mK, the constant $\beta_{Kn} = 1.5$, the diameter of the gas molecules $d_g = 3.76 \cdot 10^{-10}$ m and $T = 293.15$ K (compare [12]).

In case of vacuum insulation panels (VIP), an open-cellular porous materials combined with a low gas pressure and surrounded by a gas-tight envelope (often made of aluminium) used to prevent air from entering the panel [12]. For instance the thermal conductivity of fumed silica $\lambda_{therm} \approx 0.020$ W/mK can be lowered to $\lambda_{therm} \approx 0.003$ W/mK by applying a vacuum of $p_g \approx 5 \cdot 10^3$ Pa. However, VIPs are expensive, very fragile, sensitive towards puncture and difficult to process. Furthermore, permeation with time cannot be entirely excluded, which results in an increasing thermal conductivity. Thus, in order to achieve a highly efficient thermal insulation material, a reduction of the pore size is the most promising method.

The best-known example for nanoporous bulk materials are aerogels. Organic, inorganic as well as organic-inorganic hybrid materials are already described in literature [4, 16–18]. The preparation of aerogels via the sol-gel-process was first described in 1931 by *Kistler* [4, 5]. The first step is the formation of sol particles with a diameter of 1–3 nm from dissolved molecular precursors [19]. Condensation of these sol particles forms a spongelike, three-dimensional gel network. By definition, in the case of aerogels, the solvent is replaced by air. Though, conventional drying by evaporating the solvent results in shrinkage of the material due to capillary forces. To avoid altering of the gel network, a supercritical (sc) drying step using e.g. CO_2 is necessary. However, in the case that water is used as solvent, firstly water has to be replaced by e.g. acetone or ethanol. Subsequently the liquid can be exchanged by $scCO_2$, whereby the required time is connected to the diffusion of $scCO_2$ and thus the dimensions of the gel. E.g. in case of rods with a diameter of 1.5 cm, the replacement takes 3 hours [20]. In sum, the production of aerogels requires an autoclave and is not readily possible in a continuous way. Furthermore, a solvent-replacement is required, which is time-consuming as well as cost-intensive.

Another approach for the production of nanoporous bulk materials is the "Nanofoams by Continuity-Inversion of Dispersion" (NF-CID) procedure developed by *Strey* and *Müller* [21, 22]. Here, in a first step a dispersion of monodisperse, thermoplastic polymeric nanoparticles, e.g. PS or polymethylmethacrylate (PMMA) is prepared via emulsion polymerization. Drying yields a close-packed colloidal crystal, whose nanodisperse voids in the next step are filled with a supercritical fluid, e.g. $scCO_2$. Thereby the supercritical fluid reduces the glass transition temperature T_g of the polymer [23]. Increasing the temperature above the lowered T_g results in the continuity-inversion, i.e. the colloidal crystal is converted into a polymer matrix with nanometer sized spherical pools of the supercritical fluid. Expansion results in a foaming and a simultaneous fixation of the polymer, since the glass transition is increased again due to the decreasing amount of sc-fluid soluble in the polymer matrix. It is found that the foam pore size is directly proportional to the particle size [22]. In addition, the use of a colloidal crystal template made of polymeric nanoparticles allows a rapid saturation of the thermoplastic polymer with the supercritical blowing agent and thus a rapid decrease of T_g. Moreover, the colloidal crystal template provides the high number density of blowing agent pools. Applying the NF-CID procedure, PS and PMMA foams with pore diameters in the range of 0.1 to 10 µm were realized [22, 24].

Very recently, *Grassberger et al.* described a blowing-agent free process to produce a nanostructured polymer material [24–26]. A polymeric network is swollen by a homogeneous mixture of two solvents. While the first solvent decreases T_g and swells the polymer, the second solvent does not swell the polymer but must have a lower vapor pressure than the first one. PMMA for instance can be swollen with a mixture of acetone and cyclohexane acting as swelling and non-swelling agent, respectively. A slight volume decrease occurs upon evaporation of the swelling-agent, while the network cannot collapse due to the embedded second solvent. Furthermore, T_g of the polymer is increased and the fixation is achieved. At the end the non-swelling agent can then be evaporated to obtain the nanostructured polymer material. Thus, PMMA materials with a pore diameter between 80 and 800 nm were realized [25]. Furthermore, *Grassberger et al.* showed that the gaseous thermal conductivity of these PMMA materials can be described by the *Knudsen*-effect. By decreasing the characteristic pore size, the gaseous thermal conductivity at ambient pressure was reduced from $\lambda_g = 0.026$ W/mK for 300 µm down to $\lambda_g = 0.009$ W/mK for 80 nm [25].

However, the procedures for the production of nanoporous materials presented so far imply several elaborate steps, e.g. require the synthesis of a dispersion of monodisperse, thermoplastic polymeric nanoparticles or polymer networks in the first step. In addition often different solvents are required. In contrast the POSME principle described below, requires only one blowing agent and should generally facilitate a continuous production of polymeric nanofoams.

The "Principle Of Supercritical Microemulsion Expansion" (POSME) was proposed by *Strey, Sottmann* and *Schwan* in 2003 [27, 28]. This approach uses thermodynamically stable and nanostructured microemulsions as the templates for the production of highly porous nanostructured foams. Microemulsions are mixtures of at least three components, whereby an amphiphilic film facilitates the mixing of two otherwise immiscible components, i.e. a hydrophilic and a hydrophobic component. In case of the POSME approach, the replacement of the hydrophobic component with a near or supercritical fluid, e.g. scCO$_2$, allows for the formulation of a foamable nanostructured template consisting of supercritical fluid-swollen domains in an hydrophilic matrix with number densities up to 10^{19} cm^{-3} [28]. Thereby CO$_2$ is the supercritical fluid of choice, because it is abundant, inflammable, non-toxic and inexpensive. Furthermore CO$_2$, exhibits a relatively high critical density of $\rho_c = 0.468$ g·cm^{-3} at technically feasible critical pressure $p_c = 73.77$ bar and temperature $T_c = 30.98$ °C [29]. Adjusting a temperature above T_c, the density of the supercritical fluid reduces steadily and without a nucleation step upon expansion of the template microemulsion. Thus, ideally each CO$_2$-swollen micelle will be transferred into a foam bubble. However, a precise knowledge of the phase behavior and microstructure of these template microemulsions is essential to apply the POSME procedure. First experiments showed that both the expansion step and the solidification of the liquid foam is challenging. *Klostermann* showed that sugar foams with a pore diameter of 1– 30 µm and a nano-sized substructure can be produced by foaming microemulsions consisting of highly concentrated sugar solutions and propane [30, 31]. The coarsening of the structure was explained by *Ostwald* ripening [32–34] and coagulation followed by coalescence [34–36].

In 2005 the Bayer MaterialScience AG (BMS) started the cooperation with the *Strey* group (Institute of Physical Chemistry, University of Cologne), which aimed at the production of nanocellular polyurethane foams as a highly efficient thermal insulation material utilizing the POSME procedure (Figure 1.2).

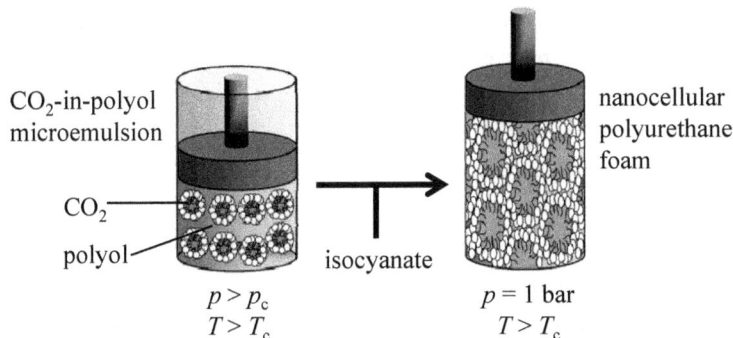

CO_2-in-polyol microemulsion

CO_2

polyol

isocyanate

$p > p_c$
$T > T_c$

nanocellular polyurethane foam

$p = 1$ bar
$T > T_c$

Figure 1.2: Schematic representation of the most relevant steps of the "Principle Of Supercritical Microemulsion Expansion" (POSME) process for the production of a nanocellular polyurethane (PU) foam. First a microemulsion of CO_2-swollen micelles in a polyol matrix is formulated at supercritical conditions ($p > p_c$, $T > T_c$). After addition of polyisocyanate, which starts the polyaddition reaction, the expansion to ambient pressure induces the foaming of the solidifying foam. Modified from [28].

Within the project polyol-CO_2-microemulsions were formulated at the University of Cologne and afterwards foamed at BMS. As has been found for the sugar-foams, the PU foams produced by the POSME principle show coarsening phenomena, which are partly related to the high monomeric solubility of CO_2 in polyol. In order to reduce these coarsening phenomena, the Anti-Aging-Agent-concept was proposed [37–39]. In this approach a low-molecular, hydrophobic co-oil is added as an Anti-Aging-Agent (AAA) to decrease the unfavorable interfacial tension between CO_2 and polyol during the expansion step. In mid-2015 a cooperation between BMS and the *Sottmann* group (Institute of Physical Chemistry, University of Stuttgart) was established, which was followed by the project "Polyurethane nano-cellular foams made from blowing agent based microemulsions for high performance thermal insulation" (PuNaMi) funded by the German Federal Ministry for Economic Affairs and Energy (BMWI). Note that BMS has become a separate legal entity in 09/2016, operating under the name Covestro Deutschland AG. Affiliated partners are the Covestro Deutschland AG, Evonik Industries AG, CAM-D, the Fraunhofer LBF, as well as groups from the Johannes Gutenberg University Mainz and Friedrich Alexander University Erlangen-Nürnberg. Thus, the effort and focus to use the POSME procedure for the production of nano-cellular PU foams were increased.

As pointed out, the detailed knowledge of the phase behavior and microstructure of these polyol-containing CO_2-microemulsions, as well as the influence of the expansion step on the microemulsion properties is crucial to put the POSME procedure into practice. Thus, the formulation and characterization of well-structured polyol-rich CO_2-microemulsions exhibiting a high number density of CO_2-swollen micelles was the major task of the subproject of the University of Stuttgart.

In order to accomplish this task, the detailed knowledge of the properties of closely related microemulsion systems of the type water – CO_2 – non-ionic surfactant might be of great benefit. In recent years these systems have also attracted increasing attention due to their potential as a possible replacement for organic solvents and as reaction media [40]. Furthermore, they are also highly interesting from a fundamental point of view, since their properties such as bending elasticity constants can strongly be influenced by varying the pressure without changing the composition [41]. Unfortunately, environmentally unfriendly fluorinated surfactants are the amphiphiles of choice to solubilize CO_2 in water efficiently [42–44]. In recent years a number of aqueous CO_2-microemulsion systems have been systematically studied with respect to their phase behavior, microstructure and the properties of the amphiphilic film [41, 45–51]. Nevertheless, so far the entire phase inversion and the respective structural inversion starting from CO_2-in-water over bicontinuously structured to water-in-CO_2 microemulsions induced by temperature wasn't shown.

Similar to the corresponding state description of real gases by *van der Waals* in 1881 [52], *Strey* and *Sottmann* have shown that for microemulsion systems of the type water – *n*-alkane – C_iE_j the temperature dependence of the phase behavior, the oil/water-interfacial tension σ_{ab} and the characteristic length scale ξ can be scaled onto each other [53–56]. However, until now it was not proven, whether a corresponding state description also exists for water – CO_2 – non-ionic surfactant microemulsion systems.

Recently, for the microemulsion system of the type water – CO_2/cyclohexane – fluorinated surfactant *Pütz et al.* observed that the efficiency follows a pressure-specific parabolic trajectory as a function of the mass fraction of the co-oil cyclohexane [47, 57]. Thereby, the efficiency is maximal at a certain cyclohexane mass fraction and a reduction of the amount of surfactant needed to formulate a one-phase microemulsion by a factor of 2 to 5 was found. From SANS-measurements an accumulation of cyclohexane in the center of the CO_2-swollen micelles due to repulsive interactions between the cyclohexane and the fluorinated hydrophobic surfactant part was found [47]. Moreover, it was proposed that the increase of the efficiency is connected to an increase of the bending rigidity of the amphiphilic film, however, a probe is so far missing [58, 59].

Objectives

The first goal of this work was a fundamental study of model-type aqueous CO_2-microemulsions stabilized by the new fluorinated, environmental friendly surfactant Capstone® FS-3100. A second task was to use the results obtained for the model-type aqueous CO_2-microemulsions to formulate and characterize well-structured polyol-rich CO_2-microemulsions as templates for the production of nano-cellular polyurethane-foams via the POSME principle.

Thus, the main aim of the first part was to show, that model type aqueous CO_2-microemulsions stabilized by the nonionic Capstone® FS-3100 resemble the main features of classical water/oil-microemulsions stabilized by n-alkyl polyglycol ether surfactants [53, 54, 56]. Consequently, the phase behavior of the water-rich, balanced as well as CO_2-rich microemulsions had to be characterized firstly by recording the respective sections through the phase prism at different pressures between 150 and 300 bar. With the knowledge of the phase diagrams, a systematic study of the microstructural evolution by SANS was conducted in order to prove whether a temperature dependent scaling of the characteristic length scale is possible. Furthermore, additional NSE measurements should provide the pressure dependence of the bending rigidity of the amphiphilic film. Having in mind the use of co-oils as so-called Anti-Aging-Agents (AAAs), the influence of the co-oil cyclohexane on the phase behavior, microstructure and dynamics of balanced aqueous CO_2-microemulsions had to be studied. Main focus was, to prove whether the partial replacement of CO_2 by cyclohexane leads to an increase of the bending rigidity, as predicted in studies of aqueous CO_2-microemulsions stabilized by Zonyl® surfactants [47].

As the production of nano-cellular polyurethane-foams via the POSME principle was the main driving force of this thesis, the goal of the second part was the formulation and characterization of well-structured polyol-rich CO_2-microemulsions using the results obtained for the model-type aqueous CO_2-microemulsion. In order clarify, whether nonionic trisiloxane surfactants are able to form micelles in polyol, measurements of the Gibbs adsorption isotherm as well as SANS studies were planned. Subsequently, in order to formulate foamable polyol-rich microemulsions, the influence of the amount of CO_2 and pressure on the phase behavior had to be studied. Whether the added CO_2 leads to the swelling of the micelles should be proven by SANS increasing the scattering contrast via the use of a partly-deuterated polyol. Additionally, time-resolved SANS measurements were foreseen to study the structural changes due to periodic pressure jumps. Finally 1,2 and 1,10-decanediol were planned to be added as AAAs in order to study their influence on the phase behavior, microstructure and demixing kinetics.

2 Fundamentals

2.1 Microemulsions

Microemulsions are defined as thermodynamically stable, macroscopically isotropic, but nanostructured mixtures of at least three components, stabilized by an amphiphilic film [54, 56, 60, 61]. The first microemulsion system water – oil – alkali-metal soap was already described in 1943 by *Hoar* and *Schulman* [62]. The term microemulsion, however, was first used by Schulman in 1959 [63]. The first two components, typically a polar (A, e.g. water) and a non-polar component (B, e.g. an oil), are mutually immiscible. The third component is an amphiphile (C, surfactant). It allows for the formulation of a microemulsion by solubilizing both A and B completely within each other, whereas one-phase results. The surfactant, which consists of a polar (lipophobic) as well as a non-polar (hydrophobic) molecular part, adsorb due to this amphiphilicity at the interface between water and oil. Thereby these phases are separated by the formed amphiphilic film. This process decreases the interfacial tension σ_{AB} between A and B effectively to almost zero, while at the same time the interfacial area strongly increases. As result of the low interfacial tension the surface energy E_S is raised by the thermal energy k_BT, which is why these systems are thermodynamically stable. Even though the systems are macroscopically homogeneous, the two immiscible components are still separated from each other via the amphiphilic film. Depending on the composition of the system, temperature T and pressure p, different structures like layers (e.g. lamellar L_α-phase) or droplets are formed. The influence of temperature on the phase behavior of ternary systems was intensively studied by *Kunieda, Shinoda* and *Friberg* [64, 65]. In 1987 *Gale et al.* were the first to study pressure-dependent microemulsions containing a supercritical fluid as the hydrophobic component [66].

2.1.1 Phase behavior

As pointed out above, microemulsions consist of at least three components, while being stabilized by a surfactant. In general cationic, anionic, zwitterionic and non-ionic surfactants can be used for the formulation of microemulsion systems. Since only non-ionic surfactants were used in this work, the following section is limited to microemulsions stabilized by non-ionic surfactants. As shown by *Kahlweit et al.*, at a constant pressure, the phase behavior of ternary systems of the type water (A) – oil (B) – non-ionic surfactant (C) can be best shown in a temperature dependent *Gibbs* phase triangle, since the phase behavior is rather complex [67]. For a better understanding of the complex phase behavior of ternary systems, looking at the three temperature-dependent binary side systems of the type water – oil (A – B), oil – non-ionic surfactant (B – C) and water – non-ionic surfactant (A – C) can be helpful.

Binary side systems

As shown in Figure 2.1, the binary side system of the type water – oil (A – B) represents the simplest one. Water and oil are mutually immiscible with each other over the entire temperature range. Consequently a pronounced lower miscibility gap with an upper critical point, normally above 100 °C, is found. In contrast, the binary side system of the type oil – non-ionic surfactant (B – C) exhibits a lower miscibility with an upper critical point cp_α. The critical temperature T_α is in the range of the melting point of the mixture, whereas it is strongly dependent on the nature of both oil and non-ionic surfactant [60].

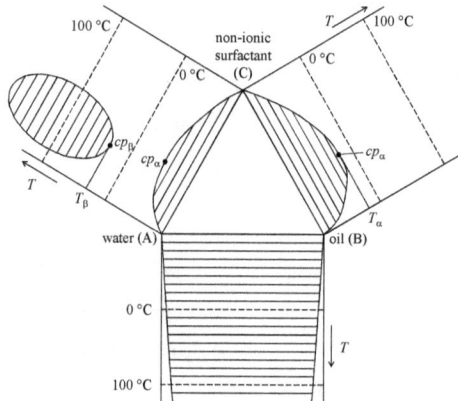

Figure 2.1: Schematic drawing of the three binary side systems of water (A), oil (B) and non-ionic surfactant (C), whose superposition describes the ternary system. The components A and B are mutually immiscible, while the lower miscibility gap of the binary system B – C is often located below 0 °C. The binary system A – C shows a lower miscibility gap at low temperatures and an upper miscibility gap, which is relevant for the description of microemulsions. Redrawn from [60].

The more complex binary side system water – non-ionic surfactant (A – C) shows a lower as well as an upper miscibility gap. The lower miscibility gap with the upper critical point cp_α lies in most cases far below the melting point of the mixture and can therefore be neglected. The upper miscibility gap with the lower critical point cp_β, on the other hand, is located in the range of $T = 0$–100 °C and is highly relevant for the microemulsion system. Again, the corresponding temperature T_β is strongly dependent on the surfactant used. The upper critical point of the upper miscibility gap lies far above the boiling point of the mixture and is therefore not relevant for further considerations. At intermediate temperatures and surfactant mass fractions below the critical micelle concentration (*cmc*) a homogeneous mixture of water and non-ionic surfactant can be found. At surfactant concentrations above the *cmc* the surfactant starts to self-agglomerate in water. Thus, micelles are initially formed, and at even higher surfactant concentrations higher ordered liquid crystalline phases appear [68].

Ternary systems

At constant temperature, the ternary system can be described by the *Gibbs* phase triangle. The temperature dependent *Gibbs* phase triangles of the system water (A) – oil (B) – non-ionic surfactant (C) forming a *Gibbs* phase prism are shown in Figure 2.2. The *Gibbs* phase prism can be obtained from the interplay of the three binary side systems (compare Figure 2.1). Thereby the complex phase behavior is mainly a result of the overlap of the lower miscibility gap of the oil – non-ionic surfactant (B – C) system with the upper miscibility gap of the water – non-ionic surfactant (A – C) system. Furthermore, the mutual immiscibility of water and oil is crucial. Consequently, the non-ionic surfactant becomes better oil-soluble with increasing temperature. The decreasing water- and increasing oil-solubility of the surfactant with increasing temperature is the main cause for the strong temperature-dependency of the phase behavior of ternary systems. This is due to an ongoing weakening of the hydrogen bonds between the water molecules and the ethoxy units of the surfactant with increasing temperature [69].

In general small amounts of surfactant molecules are monomerically dissolved in water as well as in oil [70]. Starting at temperatures below the lower temperature T_l, pronounced hydrogen bonds between water and surfactant and weak interactions between oil and surfactant can be found. Accordingly, a surfactant-rich oil-in-water (o/w) microemulsion coexists with an oil excess phase. This state is denoted as $\underline{2}$ to indicate the surfactant-rich phase, which in this case is the lower one. By increasing the temperature above T_l, the hydrogen bonds between water and surfactant are weakened and the solubility of surfactant molecules in water decreases steadily. Furthermore the solubility of surfactant in oil increases with temperature. Thus, the surfactant-rich o/w-microemulsion separates at T_l into a surfactant-rich middle phase and a water excess phase. Consequently three phases (a water excess phase, a surfactant-rich microemulsion phase and an oil excess phase) coexist, termed 3. By further increase of the temperature the surfactant becomes better soluble in oil than in water and the composition of the oil excess phase and middle phase converges. At the upper temperature T_u these compositions are equal and the phases merge to form a surfactant-rich water-in-oil (w/o) microemulsion and a coexisting water excess phase. Since the density of the surfactant-rich phase is normally lower than the one of the water excess phase this state is denoted as $\overline{2}$.

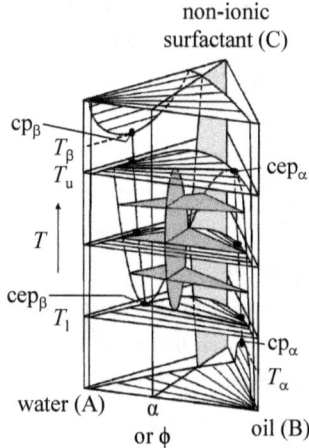

Figure 2.2: Schematic *Gibbs* phase prism of the system water (A) – oil (B) – non-ionic surfactant (C). The phase prism can be obtained by stacking isothermic phase triangles. Additional, a so called $T(\gamma)$-section, i.e. phase behavior as a function of temperature and overall surfactant concentration, at a constant water to oil ratio α/ϕ is drawn in. The three-phase region is colored in grey, while the one-phase region is drawn in blue. [70] – Reproduced by permission of The Royal Society of Chemistry

$T(\gamma)$-section

Since temperature-dependent *Gibbs* phase triangle are quite complex, the determination of all phase boundaries is very time-consuming. However, earlier studies have shown that different sections through the phase prism are suitable for describing the phase behavior of ternary systems [56]. One convenient section is the so-called $T(\gamma)$-section at a constant oil mass fraction α in the oil/water mixture

$$\alpha = \frac{m_B}{m_A + m_B}$$

2.1

with m_i the mass of component i.

Thus, the phase behavior is determined as a function of the temperature and the overall surfactant mass fraction

$$\gamma = \frac{m_C}{m_A + m_B + m_C} \qquad\qquad 2.2$$

Alternatively, for microemulsion systems with known densities of each component, the volume fraction of oil in the oil/water-mixture ϕ and the overall surfactant volume fraction ϕ_C can be used to describe the respective sample composition.

$$\phi = \frac{V_B}{V_A + V_B} \qquad\qquad 2.3$$

$$\phi_C = \frac{V_C}{V_A + V_B + V_C} \qquad\qquad 2.4$$

Here V_i being the volume of component i.

With respect to the study of the pressure- and temperature-dependent phase behaviors of CO_2-microemulsion systems, which is the main topic of this work, the use of mass fractions is more appropriate.

An example of a $T(\gamma)$-section, and the present phases, is shown schematically in Figure 2.3.

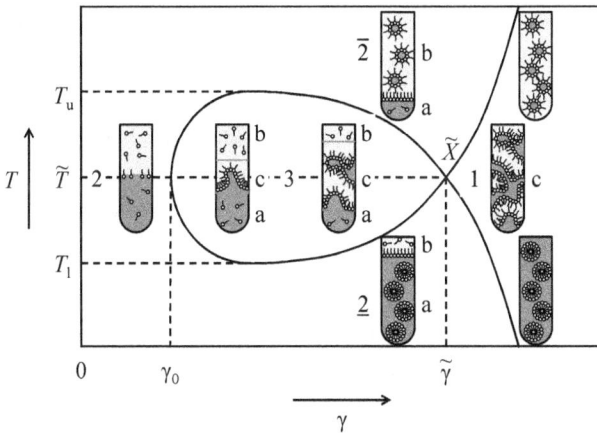

Figure 2.3: Schematic $T(\gamma)$-section of the microemulsion system water (A) – oil (B) – non-ionic surfactant (C) at constant oil mass fraction α. The test tubes exemplarily show the structures of the present phase states. At surfactant mass fractions below γ_0 the surfactant is either dissolved monomerically in both phases (A/B) or absorbed at the macroscopic interface. At $\gamma_0 < \gamma < \tilde{\gamma}$ the phase sequence $\underline{2} \rightarrow 3 \rightarrow \overline{2}$ occurs, while at higher surfactant mass fractions a one-phase region instead of a three phase region can be found. Redrawn from [56].

At surfactant mass fractions below γ_0 a water and an oil phase coexist over the entire temperature range. Thereby the surfactant is monomerically dissolved in both phases and partially absorbed at the macroscopic interface. At γ_0 both phases as well as the macroscopic interface are saturated with surfactant. Thus, increasing the surfactant mass fraction to $\gamma_0 < \gamma < \tilde{\gamma}$ the phase sequence $\underline{2} \to 3 \to \overline{2}$ can be found with increasing temperature. At low temperatures the surfactant preferentially dissolves in water. Thus, a surfactant-rich o/w-microemulsion coexists with an oil excess phase ($\underline{2}$). In contrast, at high temperatures the surfactant is better soluble in oil and a surfactant-rich w/o-microemulsion coexists with a water excess phase ($\overline{2}$). By increasing the surfactant mass fraction from γ_0 to $\tilde{\gamma}$ the volume of both excess phases decreases in favor of the surfactant-rich middle phase, until finally at the so-called optimum point $\tilde{X}$ the excess phases merge with the middle phase to build up the one-phase region. $\tilde{\gamma}$ defines the minimum amount of surfactant that is sufficient to solubilize given amounts of water and oil within one-phase. The formed microemulsion phase is termed as 1. At lower or higher temperatures the two phase states $\underline{2}$ and $\overline{2}$ are found, respectively. By increasing the surfactant mass fraction the one-phase region increases, while at even higher surfactant mass fractions higher ordered liquid crystalline phases like L_α can be formed. Note that for symmetric $T(\gamma)$-sections the mean temperature $T_m = (T_u + T_l)/2$ equals the temperature $\tilde{T}$.

Water- and oil-rich microemulsions

The common $T(\gamma)$-section can be used to determine the minimal amount of surfactant $\tilde{\gamma}$ that is sufficient to solubilize given amounts of water and oil at constant α. Though, for systems with only small amounts of oil or water the $T(w_B)$- and $T(w_A)$-sections with constant values of γ_a or γ_b are preferable [56]. Water-rich microemulsions are investigated at a constant mass fraction γ_a of surfactant in the water-surfactant mixture

$$\gamma_a = \frac{m_C}{m_A + m_C} \qquad\qquad 2.5$$

The phase behavior is then examined as function of temperature and overall oil mass fraction

$$w_B = \frac{m_B}{m_A + m_B + m_C} \qquad\qquad 2.6$$

For oil-rich microemulsions the amount of added water is specified by the overall water mass fraction

$$w_A = \frac{m_A}{m_A + m_B + m_C} \qquad 2.7$$

to the binary oil-surfactant mixture, which is defined by the mass fraction

$$\gamma_b = \frac{m_C}{m_B + m_C} . \qquad 2.8$$

Figure 2.4 schematically shows a temperature-dependent *Gibbs* phase triangle with marked $T(w_B)$- and $T(w_A)$-sections. The phase sequence of the $T(w_B)$- and $T(w_A)$-sections are analogous to that of the $T(\gamma)$-section.[71]

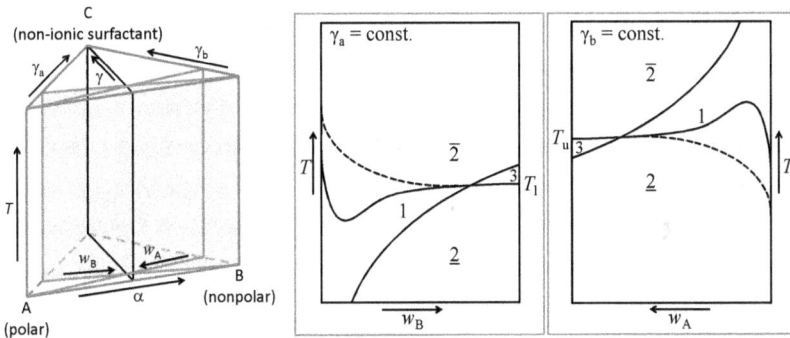

Figure 2.4: Phase prism (left) with highlighted sections. The $T(\gamma)$-section is again drawn in (black). $T(w_B)$-section (orange, middle) as a function of temperature T and overall oil mass fraction w_B at a constant mass fraction of surfactant in the binary water – non-ionic surfactant-mixture γ_a. Please note that the upper phase boundary is shown for "strong" (solid line) as well as for "weak" amphiphiles (dashed line). $T(w_A)$-section (blue, right) as a function of temperature T and overall water mass fraction w_A at a constant mass fraction of surfactant in the binary oil – non-ionic surfactant-mixture γ_b again for "strong" as well as "weak" amphiphiles. In general the phase sequence $\underline{2} \rightarrow 1 \rightarrow \overline{2}$ or $\underline{2} \rightarrow 3 \rightarrow \overline{2}$ can be found with increasing temperature. [54, 71, 72]

As can be seen, for the $T(w_B)$-section (Figure 2.4, middle) at low temperatures and small oil mass fractions w_B a one-phase region can be found, which results from the good solubility of surfactant in water at these temperatures. At elevated temperatures a two-phase coexistence of a water-in-oil microemulsion phase with a water excess phase can be found ($\overline{2}$). In case of "strong" amphiphiles (solid line in Figure 2.4, middle) the upper *near-critical boundary* (*ncb*) runs through a pronounced minimum [56, 73]. In contrast a monotonically decay is observed for "weak" amphiphiles (dashed line). The *ncb* starts at the upper miscibility gap of the binary system water – non-ionic surfactant. The minimum results from a so-called closed loop in the *Gibbs* phase triangle. The lower so-called *oil-emulsification failure boundary* (*oefb*) [74, 75] ascends monotonically and indicates the ability to solubilize oil in the given water – non-ionic surfactant mixture. Both phase boundaries intersect at T_l, where the one-phase region closes. By increasing the oil mass fraction w_B a microemulsion phase coexists with an oil excess phase.

The $T(w_A)$-section shows an inverse but analogous phase behavior (Figure 2.4, right). Here the upper *water emulsification failure boundary* (*wefb*) defines the ability to solubilize water in a given oil – non-ionic surfactant mixture. The *wefb* intersects with the lower *near-critical boundary* (*ncb*) at T_u. The *ncb* of the $T(w_A)$-section starts from the lower miscibility gap of the binary system oil – non-ionic surfactant. Again, the *ncb* of systems with "weak" amphiphiles ascends monotonically, while a maximum is observed for "strong" amphiphiles.

Increasing the amount of surfactant γ_a or γ_b enables to solubilize larger amounts of oil or water, respectively. Thus the emulsification failure boundary shifts to higher oil or water mass fractions.

The optimum point

The location of the optimum point $\tilde{X}$ defined by $\tilde{\gamma}$ and $\tilde{T}$ is unique for each microemulsion system and depends on e.g. the hydrophobicity of the oil used. Thereby values of $\tilde{\gamma}$ as small as possible are desirable for technical applications. In addition, one parameter that strongly influences the location of the $\tilde{X}$-point is the water-to-oil ratio (α or ϕ). Figure 2.5 shows a ϕ-variation and the resulting $T(\gamma)$-sections of the microemulsion system $H_2O - n$-octane $- C_{12}E_5$. $C_{12}E_5$ is a n-alkyl polyglycol ether (C_iE_j surfactant), in which i defines the number of carbon atoms in the hydrophobic chain and j the number of hydrophilic ethoxy units. C_iE_j surfactant are frequently used non-ionic surfactants, as i and j can be varied independently.

As shown in Figure 2.5, the efficiency of the system with $\phi = 0.6$ is lowest, but increases by increasing or decreasing the water-to-oil ratio ϕ. In either case, the volume fraction of the component to be solubilized by the surfactant decreases, hence less surfactant is needed. Thereby the $\tilde{X}$-point shifts along a parabolic trajectory. Thus, an increase of ϕ results in an increasing phase inversion temperature $\tilde{T}$ and vice versa [53, 55].

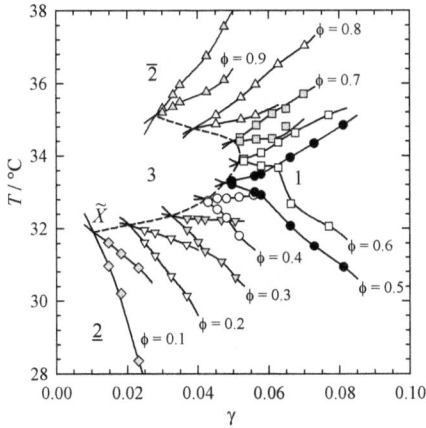

Figure 2.5: $T(\gamma)$-sections of the microemulsion system $H_2O - n$-octane $- C_{12}E_5$ with varying water-to-oil ratio of $\phi = 0.1$ to 0.9. The temperature $\tilde{T}$ at the optimum point $\tilde{X}$ increases with increasing oil content ϕ, while $\tilde{\gamma}$ runs through a maximum at $\phi \approx 0.6$. Data taken from [76].

Furthermore, the location of the $\tilde{X}$-point is dependent on the amphiphilic properties of the surfactant. In Figure 2.6 the influence of the hydrophilicity and lipophilicity is demonstrated by plotting the temperature $\tilde{T}$ at the $\tilde{X}$-point of the microemulsion system $H_2O - n$-octane $- C_iE_j$ ($\phi = 0.50$) as a function of the surfactant mass fraction $\tilde{\gamma}$ at the optimum point $\tilde{X}$ for different surfactants with (i,j).

As can be seen in Figure 2.6, increasing the hydrophobic chain length from $i = 6$ to 12 increases the solubilisation efficiency of the surfactant and therefore $\tilde{\gamma}$ decreases to smaller values. Furthermore, the phase inversion temperature $\tilde{T}$ decreases slightly. By increasing the number of hydrophilic ethoxy units j the phase inversion temperature $\tilde{T}$ increases, accompanied by a slight decrease of the surfactant mass fraction $\tilde{\gamma}$ [70]. Thus the efficiency of ternary systems can be easily adjusted by the variation of the number of carbon atoms in the hydrophobic surfactant chain i and the number of ethoxy units j.

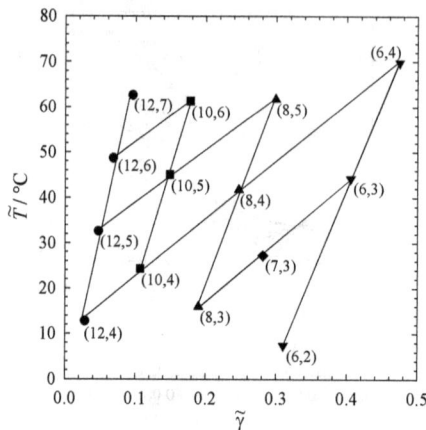

Figure 2.6: Efficiency points $\tilde{X}$ of the microemulsion system $H_2O - n\text{-octane} - C_iE_j$ at $\phi = 0.50$ as function of $\tilde{\gamma}$ and $\tilde{T}$. (i,j) indicates the individual values for i and j. Increasing the number of carbon atoms i mainly decreases $\tilde{\gamma}$, while an increasing number of hydrophilic ethoxy units j mainly cause an increase of the phase inversion temperature $\tilde{T}$. Redrawn from [70].

Technical grade surfactants

So far, the phase behavior of microemulsions containing pure surfactants was described. However, the synthesis of pure surfactant is very expensive and the usage in large-scale applications therefore uneconomical. Therefore, the use of technical grade C_iE_j surfactants is preferred. n-Alkyl polyglycolether surfactants are synthesized by ethoxylation of the respective fatty alcohols. In case of technical grade surfactants the ethoxylation degree exhibits a broad distribution. Furthermore, non-reacted alcohol and synthesis-residues are found in the product.

Consequently, the phase behavior of microemulsions containing technical grade surfactants differs somewhat from those with pure surfactants. Figure 2.7 shows the effect of technical grade surfactants upon the phase behavior of microemulsions of the type $H_2O - n\text{-octane} - $ non-ionic surfactant. The phase behavior is shown for the microemulsion system containing pure $C_{12}E_6$ and the technical grade analogue DA-6.

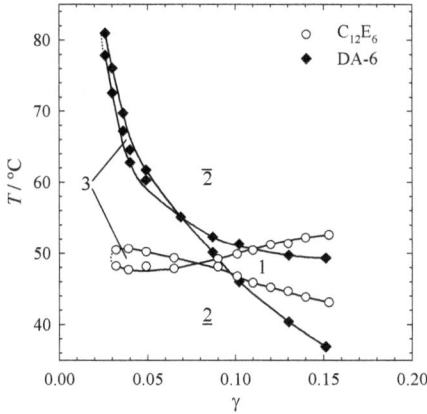

Figure 2.7: $T(\gamma)$-sections of the microemulsion system $H_2O - n$-octane – non-ionic surfactant with pure $C_{12}E_6$ (hollow circles) and the technical analogue DA-6 (black diamonds) at $\phi = 0.50$. The phase boundaries of the pure system are symmetrically located around the phase inversion temperature $\tilde{T}$. Due to the broad distribution of the ethoxylation degree, the phase boundaries of the technical system steadily shift to higher temperatures. Data taken from [77, 78].

Ternary systems stabilized by a pure C_iE_j surfactant typically show a $T(\gamma)$-section which is symmetrically located around the phase inversion temperature $\tilde{T}$ (compare Figure 2.7, hollow circles). The corresponding systems with technical grade surfactant (black diamonds), in contrast, show a distorted $T(\gamma)$-section. By decreasing the surfactant mass fraction the phase boundaries steadily shift to higher temperatures. Moreover, the system containing the technical grade surfactant is somewhat more efficient. This finding can be explained by the broad ethoxylation degree of the surfactant used and non-reacted alcohol. Surfactant molecules with lower ethoxylation degree and especially the non-reacted alcohol exhibit a higher solubility in oil than the surfactant molecules with higher ethoxylation degree. Thus, these molecules are extracted into the oil. Upon addition of water and oil, which corresponds to a decrease in γ, the amount of surfactant molecules with low and intermediate ethoxylation degree extracted into the oil more and more increases. As the surfactant molecules with higher ethoxylation degree are remaining at the interface, the amphiphilic film becomes increasingly hydrophilic. According to Figure 2.6 a more hydrophilic surfactant, i.e. larger values for j, results in an increase of the phase inversion temperature $\tilde{T}$, leading to the distortion of the phase boundaries at small γ-values.[78]

Interfacial tension

As already mentioned, the formation of an amphiphilic film is crucial for the formulation of microemulsions [79, 80]. Thereby the surfactant molecules adsorb due to their amphiphilic character at the interface and thus reduce the interfacial tension between water and oil σ_{ab} [81]. This can be illustrated by looking at the binary system water – non-ionic surfactant. The surface tension of water σ_a or more precisely the interfacial tension of the water/air interface is generally ~ 72 mN·m^{-1} [80, 82]. Adding small amounts of surfactants the surfactant molecules either are monomerically solubilized in water or adsorbed at the interface between air and water. Consequently, the surface tension of water slightly decreases upon addition of surfactant (see Figure 2.8). At higher surfactant concentrations the critical micelle concentration (*cmc*) is reached. Here the monomeric solubility of surfactant in the solvent $\phi_{C,mon,A}$ is exceeded and the interface is fully covered by the surfactant. Consequently, the interfacial tension remains constant upon further addition of surfactant. Above the *cmc* the surfactant molecules start to form aggregates like micelles within the solvent used.

Below the *cmc* the interfacial surfactant concentration Γ_C can be described by the *Gibbs* equation [83]

$$\Gamma_C = -\frac{1}{2.303RT}\frac{\partial \sigma_a}{\partial \log c}\bigg|_{T,p}$$

2.9

with $\left(\partial \sigma_a / \partial \log c\right)$ being the slope and R being the gas constant ($R = 8.31451$ J K^{-1} mol^{-1}). Furthermore, the area per surfactant molecule a_C is defined as [84, 85]

$$a_C = \frac{1}{N_A \Gamma_C}$$

2.10

with N_A being the Avogadro constant ($N_A = 6.02214 \cdot 10^{23}$ mol^{-1}). The experimental determination of a_C via SANS is more demanding [86].

Below the *cmc* the surface tension σ_a of the respective mixture can be approximated by the semi-empirically *Szyszkowski*-equation [87]

$$\sigma_a = \sigma_0 - RT \cdot \Gamma_{max} \ln\left(1 + \frac{c}{\beta_\sigma}\right)$$

2.11

with σ_0 being the surface tension of the pure solvent, Γ_{max} the maximum surface excess, c the concentration and β_σ the inverse activity coefficient. Since no interactions are considered, this equation only applies to ideal solutions.

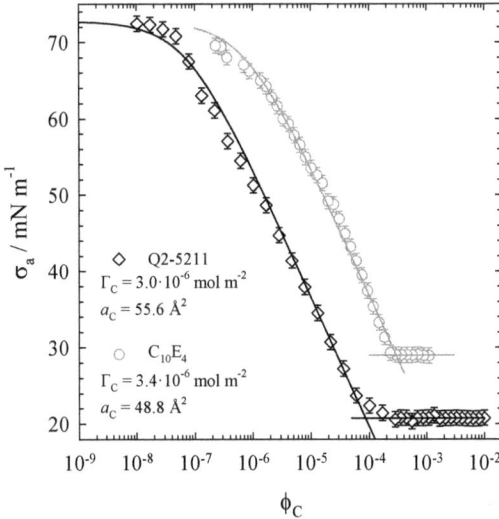

Figure 2.8: Experimentally determined surface tension of water σ_a as function of the volume fraction ϕ_C of the non-ionic surfactants $C_{10}E_4$ (circles) or Q2-5211 (diamonds, trisiloxane-based surfactant $M(D'E_{10})M$, see Figure 4.1), at $p = 1$ bar and $T = 21 \pm 1$ °C. Upon addition of surfactant the surface tension decreases from 72 mN·m^{-1} to 29.0 mN·m^{-1} for $C_{10}E_4$ and 20.7 mN·m^{-1} for Q2-5211, respectively. The characteristics of the curve can be approximated by the *Szyszkowski*-equation. By reaching the *cmc* at $\phi_C = 2.9 \cdot 10^{-4}$ ($C_{10}E_4$) and $8.7 \cdot 10^{-5}$ (Q2-5211), respectively, the surface tensions for both surfactants remains constant. [88]

2.1.2 Microstructure

As pointed out above, surfactants separate the polar from the non-polar component by forming microscopically an amphiphilic layer. While microemulsions are macroscopically homogeneous, depending on their composition, pressure and temperature, different microstructures can be found within microemulsions. The phase behavior and thus the microstructure can be described via the local curvature of the amphiphilic film. Mathematically, a curvature can be described at each point by the two principal curvatures [54, 56]

$$c_1 = \frac{1}{R_1} \quad \text{and} \quad c_2 = \frac{1}{R_2} \; . \tag{2.12}$$

Here, R_1 and R_2 are the two principal curvature radii. As can be seen in Figure 2.9 these radii can describe the curvatures of spheres as well as sponge-like structures, while in case of spherical structures $c_1 = c_2 = R_0^{-1}$ hold, with the mean radius R_0. These structures correspond to micellar and bicontinuous microemulsion structures, respectively.

a) b)

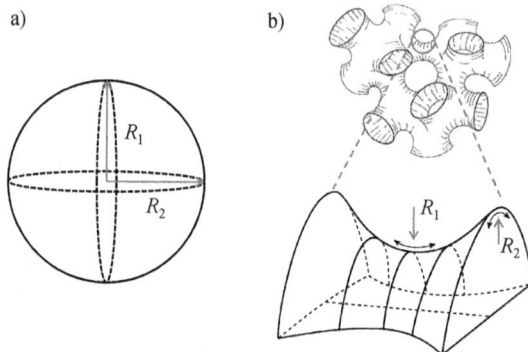

Figure 2.9: Principal curvature radii R_1 and R_2 of a) a spherical and b) a sponge-like structure. Modified from [89, 90].

By definition, the curvature c_i of the amphiphilic film is positive when curved around oil and negative when curved around water. The mean curvature H and the *Gaussian* curvature K are defined as combination of the principal curvatures c_i

$$H = \frac{1}{2}(c_1 + c_2) \quad \text{and} \quad K = c_1 c_2 \,.$$
2.13

According to *Helfrich* [91], in the absence of external forces, thermal fluctuations or conservation constrains, the mean curvature H merges into the spontaneous curvature H_0. Thus, H_0 is the main parameter for the description of the total free bending elasticity. The free energy of the amphiphilic film F_b is defined as [30, 91]

$$F_b = \int dS \left\{ \frac{\kappa}{2}(c_1 + c_2 - H_0)^2 + \overline{\kappa} \cdot K \right\}$$
2.14

where κ and $\overline{\kappa}$ are the bending rigidity, also called mean bending modulus, and saddle splay or Gauss modulus, respectively [56]. A deviation of the mean curvature of the amphiphilic film from the spontaneous curvature results in a change of the Helmholtz energy. This is expressed by the first term of eq. 2.14. The second term, with the saddle splay modulus $\overline{\kappa}$, represents the energy required for the formation of saddle-like structures, i.e. connected topologies.

Consequently, a bicontinuous structure emerges from a lamellar structure, when the renormalized saddle splay modulus tends towards zero [92, 93]. The bending rigidity κ and saddle splay modulus $\bar{\kappa}$ can be determined from neutron spin echo measurements (NSE), small angle neutron scattering (SANS) or dynamic light scattering (DLS) [94–96].

The different structures of microemulsion systems containing non-ionic surfactant molecules can be explained by variations of the mean curvature H. Thereby H can be described in good approximation via the molecular structure of the surfactant. The temperature-dependence of the molecular structure of the surfactant and thus the temperature-dependence of the amphiphilic film is shown in Figure 2.10. In case of C_iE_j surfactants, the surfactant consists of the hydrophilic ethoxy units and the hydrophobic alkyl chain. By the hydration of the ethoxy units with water a hydrate shell is formed. The ethoxy units together with the hydrate shell are the so-called hydrophilic head group of the surfactant. Assuming that the hydrophilic and hydrophobic group of the surfactant occupy equal volumes, the surfactant shows a cylindrical structure with the mean curvature $H = 0$. The surfactant, in contrast, has a cone-like shape when the two volumes are different, which is the case at higher and lower temperatures. In a 2-dimensional view this is represented by a wedge-like shape.

At low temperatures a pronounced hydrate shell can be found. Thus, the relative volume of the hydrophilic head group predominates and the amphiphilic film is curved around the oil (o/w, $H > 0$). With increasing temperature to intermediate temperatures the hydrogen bonds become weaker and the volume of the head group shrinks. In addition, the number of possible conformations of the hydrophobic alkyl chain is increasing and more oil molecules are interpenetrating into the amphiphilic film. Thus, the volume of the hydrophobic group increases with temperature. Consequently, at intermediate temperatures the surfactant exhibit equal volumes of the hydrophilic and hydrophobic group. By a further increase of temperature, the volume of the hydrophobic part of the surfactant exceeds the according volume of the hydrophilic group and a negative mean curvature can be found. Consequently, the amphiphilic film shows a temperature-dependent inversion of the mean curvature: from $H > 0$ (o/w) at low T, via a locally planar film with $H = 0$ (bicontinuous) at intermediate T to $H < 0$ (w/o) at high T.

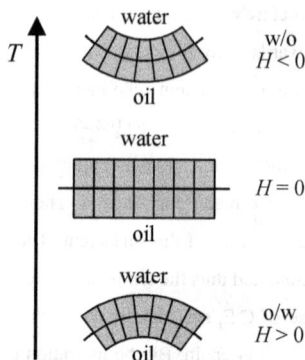

Figure 2.10: Schematic representation of the temperature-dependent mean curvature H of a non-ionic surfactant forming an amphiphilic film between water and oil. The surfactant molecules were sketched in a wedge- or rectangular-shape. With increasing temperature the hydrate shell of the hydrophilic surfactant head group shrinks due to fading hydrogen bonds, while the volume of the hydrophobic surfactant tail increases simultaneously. Consequently, by increasing the temperature the curvature of the amphiphilic film decreases, forming o/w, locally planar and w/o-structures. Redrawn from [56].

The steady transition from spherical micelles via cylindrical micelles, bicontinuous network structures and inverse cylindrical micelles to inverse spherical micelles was studied by several groups in the past. Thereby it was shown that the microstructure of microemulsions can be determined by means of electron microscopy [97, 98], small angle scattering [98–101], NMR diffusometry [98–100] and electric conductivity measurements [56].

The properties discussed so far apply to non-ionic surfactants. Ionic surfactants in contrast show the opposite behavior [56, 102]. At low temperatures the head group possesses a smaller effective volume which results in a negative mean curvature and w/o-microemulsions are formed. However, with increasing temperature the counter ion of the head group dissociates. Thus, the ionic surfactant becomes increasingly soluble in water and due to the inversion of the mean curvature o/w-microemulsions are formed at higher temperatures.

Microstructure of balanced microemulsions

The detailed knowledge about the temperature-dependent curvature of amphiphilic films build up by non-ionic surfactants enables a better understanding of the phase behavior and the observed phase inversions.

The microstructures of balanced non-ionic microemulsions in relation to the experimental determined phase behavior is shown schematically in the $T(\gamma)$-section in Figure 2.11.

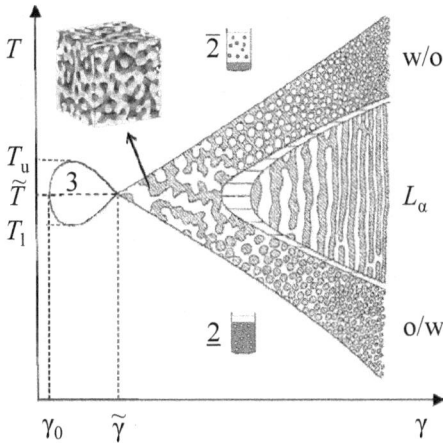

Figure 2.11: Schematic $T(\gamma)$-section of a balanced non-ionic microemulsion system with the corresponding microstructures. By increasing the temperature a phase inversion (o/w $\rightarrow$ locally planar $\rightarrow$ w/o) occurs. At $\gamma > \tilde{\gamma}$ and low temperatures an oil-in-water droplet microemulsion can be found, while at high temperatures water-in-oil droplets exist. At intermediate temperatures $T \approx \tilde{T}$ bicontinuous sponge-like or L_α-phases are formed. This is related to the inversion of the mean curvature with increasing temperature. Adapted by permission from Springer Nature [54].

At low temperatures $T \ll \tilde{T}$ the mean curvature is positive and the amphiphilic film is strongly curved around the oil. This is due to strong hydrogen bonds resulting in a large hydrate shell of the surfactant head group. As a consequence of a small curvature radius at low temperatures, the droplets can solubilize only a small amount of oil, which is expressed by a large surface-to-volume ratio. Thus, at low surfactant mass fractions γ a pronounced oil excess phase coexists with a surfactant-rich o/w-microemulsion phase. Thereby the number density of droplets at constant temperature increases steadily upon increasing the surfactant mass fraction and more oil can be solubilized. By increasing the temperature ($T < \tilde{T}$) the mean curvature decreases. Thus, the droplet radii increase and larger amounts of oil are solubilized at the respective surfactant mass fraction. Consequently, less surfactant is required to solubilize given amounts of water and oil, which explains the steady ascent of the lower phase boundary $\underline{2} \rightarrow 1$, while decreasing the value of γ.

At intermediate temperatures $T \approx \tilde{T}$ the mean curvature is $H = 0$ and the amphiphilic film is locally planar. Thus, bicontinuous sponge-like structures are formed at $\gamma > \tilde{\gamma}$. Here water- and oil-domains are macroscopically connected, while microscopically separated by the amphiphilic film. Furthermore, at high surfactant mass fractions a liquid crystalline lamellar L_α-phase may be found. The bicontinuous sponge-like structure can be described using the length scale ξ, which can be interpreted as the "mean diameter" of the formed domains. The characteristic length scale is defined as

$$\xi = \frac{d_{TS}}{2} \qquad\qquad 2.15$$

where the periodicity d_{TS} can be obtained from SANS (section 2.3.1) [54, 103]. $\tilde{\gamma}$ defines the minimal amount of surfactant needed to solubilize given amounts of water and oil, while at $\tilde{X}$ the maximal length scale ξ_m for the oil/water domains can be found. By decreasing the surfactant mass fraction the characteristic length scale increases steadily. Thereby the characteristic length scale is inversely proportional to the volume fraction of surfactant at the internal interface $\phi_{c,i}$. Consequently, the maximum length scale ξ_m is increasing with the efficiency of the surfactant, i.e. with increasing number of carbon atoms i in the hydrophobic chain of the surfactant. [56]. In contrast, increasing the number of ethoxy units j results in a decrease of the maximum length scale accompanied by an increase of the phase inversion temperature $\tilde{T}$ [86].

By increasing the temperature to $T > \tilde{T}$ the amphiphilic film starts to bend around water ($H < 0$) and water-in-oil structures are formed. Thus, at $\gamma > \tilde{\gamma}$ and above the upper phase boundary $1 \rightarrow \overline{2}$, a surfactant-rich w/o-microemulsion coexists with a water excess phase. With further increase of the temperature the curvature radius is increasing. Thus, the surface-to-volume ratio is increasing and less water can be solubilized in these droplets. Consequently, with increasing temperature increasing surfactant mass fractions are required to solubilize given amounts of water and oil. This explains the monotonic ascent of the upper phase boundary going from lower to higher γ.

Furthermore, by considering the mean curvatures also the variation of the optimum point $\tilde{X}$ upon the number of carbon atoms i in the hydrophobic chain and the number of ethoxy units j of the surfactant (Figure 2.6) can be explained. By increasing the number of carbon atoms i the volume of the hydrophobic group increase. Thus, lower temperatures are required to obtain a zero mean curvature by dehydration of the hydrophilic group. Furthermore, the bending rigidity increases, resulting in a stiffening of the amphiphilic film. Thus, less surfactant is required to solubilize given amounts of water and oil and the value of $\tilde{\gamma}$ decreases. Analogously, by increasing the number of ethoxy units j, higher temperatures are necessary to dehydrate the head group. As thermal fluctuations increase at high temperatures, the bending rigidity decreases and consequently $\bar{\gamma}$ increases. [56, 86, 92]

Microstructures of water- or oil-rich microemulsions

Starting from the considerations upon the curvature of the amphiphilic film of non-ionic surfactants, also the microstructure of water- or oil-rich microemulsion systems can be described. The schematic phase behaviors of these systems, as a function of the overall oil (left) or water (right) mass fraction, are shown in Figure 2.12.

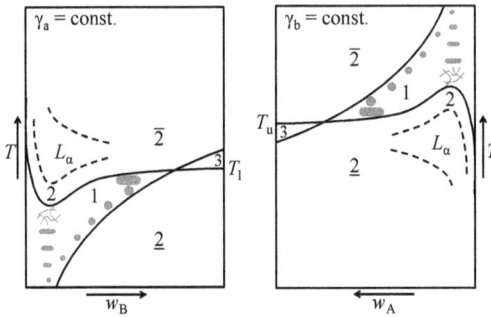

Figure 2.12: Schematic $T(w_B)$- (left) and $T(w_A)$-section (right), illustrating the microstructures that can be found in microemulsion systems using strong non-ionic amphiphiles. Close to the respective emulsification failure boundary spherical droplets can be found, whereas the radius is increasing with the amount of oil or water, respectively. By increasing or decreasing the temperature the spheres starts to elongate until cylindrical structures or even branched networks are formed. Furthermore, a lamellar L_α-phase can be found.

The $T(w_B)$- and $T(w_A)$-sections show a similar but inverse phase behavior. While in the water-rich system the oil emulsification failure boundary $oefb$ is the lower phase boundary $\underline{2} \to 1$ (Figure 2.12, left), the water emulsification failure boundary $wefb$ corresponds to the upper phase boundary $1 \to \bar{2}$ in the oil-rich system (Figure 2.12, right). Close to the respective emulsification failure boundary (efb) oil- or water-swollen droplets can be found. Here the equations $R_0 = 1/H_0$ respectively $R_0 = -1/H_0$ are valid and the droplets are maximally swollen [74]. If the temperature is increased or decreased respectively, the mean curvature H of the amphiphilic film becomes weaker and larger droplets are formed. Hence, the radius of the droplets close to the efb can be considered as the characteristic length scale of the system at the respective temperature.

The study of the microstructure of such microemulsions suggests the formation of locally cylindrical, connected structures at temperatures close to the lower ($T(w_B)$-section) and upper temperature ($T(w_A)$-section) [54, 104]. This is supported by theoretical predictions, since the formation of branched networks will be more favorable with regard to reducing the bending energy of the system [105]. At temperatures $T_l < T < T_u$ locally planar structures e.g. L_α-phase or L_3-phase (not shown in Figure 2.12) can be found [60].

Corresponding state description

In several papers it was shown, that the phase behavior, the microstructure as well as the interfacial tension of microemulsion systems of the type water – oil – non-ionic surfactant are similar to each other [60, 71, 86, 106]. Thus, in 1996, *Sottmann* and *Strey* described the so-called corresponding state of microemulsion systems by suggesting a reduced representation of the characteristic length scale and interfacial tension [53]. Here the mean temperature $T_m = (T_u + T_l)/2$, the temperature extent of the three-phase body $\Delta T = T_u - T_l$ as well as the interfacial volume fraction of surfactant at the $\tilde{X}$-point $\phi_{c,i,m}$ were described as the relevant parameters. All needed parameters can be obtained empirically from experimental data. Thus, a reduced plot of the temperature variation of the interfacial tension [53] and the characteristic length scale [55] was presented [56]. The reduced characteristic length scale $\xi \cdot \phi_{C,i,m}$ of systems of the type H_2O – n-octane – C_iE_j as a function of the reduced temperature $2(T - T_m)/(T_u - T_l)$ is shown in Figure 2.13 for the surfactants C_8E_3, $C_{10}E_4$ and $C_{12}E_5$. Note, that the length scales were determined by means of SANS as well as calculated from the compositions.

As can be seen in Figure 2.13, all data merge completely and can be described by only one renormalized empirical description according to equation 2.18 or 2.19, respectively. Furthermore, the reduced characteristic length scale $\xi \cdot \phi_{C,i,m}$ runs through a pronounced maximum at the mean temperature. Here the maximum is related to the maximum length scale ξ_m, which can be found at the $\tilde{X}$-point.

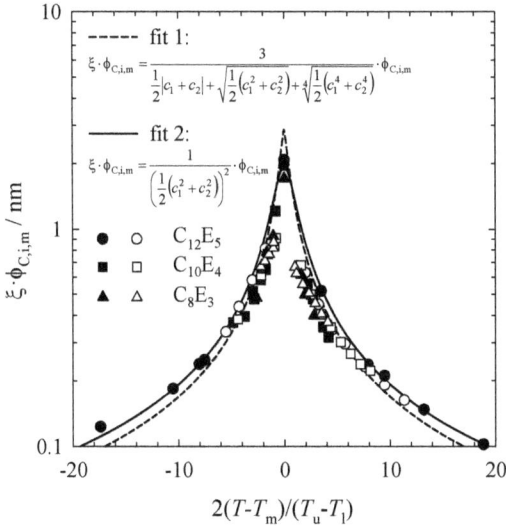

Figure 2.13: Reduced characteristic length scale $\xi \cdot \phi_{C,i,m}$ of systems of the type H_2O – n-octane – C_iE_j as a function of the reduced temperature $2(T\text{-}T_m)/(T_u\text{-}T_l)$. The length scales were determined by SANS (filled symbols) and calculated from compositions (open symbols). The fits were calculated for the system H_2O – n-octane – $C_{12}E_5$ from equations 2.18 (solid line) and 2.19 (dashed line), while the parameters where taken from [55, 86, 106]. The data was taken from [56].

According to *Strey* [54], the mean curvature H of a system can be described as

$$H = c(T_m - T)$$
2.16

while for the H_2O – n-octane – $C_{12}E_5$ system the prefactor c was found to be $c = 1.22 \cdot 10^{-3}$ Å^{-1}.

Assuming linear variations of the principal curvatures c_1 and c_2 (compare eq. 2.12)

$$c_1 = c(T_u - T) \quad \text{and} \quad c_2 = c(T_l - T),$$
2.17

the characteristic length can be defined empirically as [54]

$$\xi = \frac{1}{\left(\frac{1}{2}(c_1^2 + c_2^2)\right)^{1/2}}$$
2.18

or as [56]

$$\xi = \frac{3}{\frac{1}{2}|c_1 + c_2| + \sqrt{\frac{1}{2}(c_1^2 + c_2^2)} + \sqrt[4]{\frac{1}{2}(c_1^4 + c_2^4)}} \qquad\qquad 2.19$$

Please note that the shown curves were calculated for the system $H_2O - n$-octane $- C_{12}E_5$. The relevant parameters were determined by *Sottmann* and coworkers [55, 86, 106] to be $T_l = 31.25$ °C, $T_u = 34.45$ °C and $\phi_{C,i,m} = 0.0378$.

These findings underline the importance of the three-phase body and the $\tilde{X}$-point, as one obtain in this way the parameters T_m, ΔT and $\phi_{C,i,m}$. Consequently, the determination of these values allows suggesting the characteristic length scale ξ and interfacial tension σ_{ab}. Though, this corresponding state description was suggested for microemulsion systems of the type $H_2O -$ alkane $- C_iE_j$. A proof for the application to other microemulsion systems is missing so far.

2.1.3 scCO$_2$-microemulsions

In recent years microemulsions containing a supercritical fluid as hydrophobic component gained increasing interest due to their potential as a possible replacement for organic solvents, as reaction media or as foam template [40]. Especially near- or supercritical CO_2 ($p_c = 73.77$ bar and $T_c = 30.98$ °C) with a relatively high critical density of $\rho_c = 0.468$ g·cm^{-3} is an interesting component [29], due to the fact that CO_2 is abundant, inflammable, non-toxic and inexpensive. However, the phase behavior of supercritical fluid microemulsions is slightly different from that of microemulsions containing classical oils like e.g. n-alkanes as hydrophobic component [107].

By increasing the pressure of microemulsion systems containing n-alkanes by about $\Delta p = 100$ bar the phase behavior is slightly shifted by about $\Delta T \approx 3.5$ K to higher temperatures [108]. This is a result of the increased hydration of the surfactant head group at elevated pressures, which results in a more hydrophilic surfactant. The larger head group results in a stronger curvature of the amphiphilic film, which is why higher temperatures are required for the inversion of curvature. Hence, the phase behavior is shifted to higher temperatures. However, in case of near or supercritical fluids the stronger hydration with increasing pressure is negligible. Here, the density of such fluids is highly dependent upon temperature and pressure. In difference, alkanes show a temperature and pressure insensitive density. Since an increase in density results in stronger interactions with the surfactant molecules, an increase of the efficiency is expected and indeed provable. Furthermore, also the bending elasticity constants depends upon the pressure applied and can therefore be adjusted without changing the composition [41].

Though, due to poor interactions, most common hydrocarbon-based surfactants are not appropriate in formulating efficient CO_2-microemulsions [28, 43]. Environmentally unfriendly fluorinated surfactants, e.g. polyfluoroalkyl polyglycolether $F(CF_2)_iC_2H_4E_j$ (i is even) with a fluorinated carbon tail [109], are unfortunately most appropriate for the solubilisation of water and CO_2 [42, 43]. In recent years different types of fluorinated surfactants were synthesized and used to formulate CO_2-microemulsions [48, 110]. Thereby most studies dealt with the formulation and microstructure of water-in-CO_2 (w/c) microemulsions [48, 49, 110–113]. Further studies investigated balanced as well as CO_2-in-water (c/w) microemulsions [41, 45–47, 114]. Here especially the Zonyl® surfactants of the type $F(CF_2)_iC_2H_4E_j$ with i = 6–8 were used.

"Principle Of Supercritical Microemulsion Expansion" (POSME)

As pointed out in the Introduction, in view of the production of nanocellular materials, for e.g. thermal insulating materials [12–14], microemulsions containing a supercritical fluid, e.g. scCO_2, are promising template systems [27, 28]. Thus, *Strey, Sottmann* and *Schwan* proposed the "Principle Of Supercritical Microemulsion Expansion" (POSME) [27, 28]. Here thermodynamically stable supercritical fluid-swollen micelles with diameters of 5–20 nm are formed first. Thereby the supercritical fluid will later act as the blowing agent. Here the formulation of the microemulsion is the starting step. By performing the expansion at or above the critical temperature the density of the fluid steadily decreases from liquid-like to gas-like (Figure 2.14). Consequently, a nucleation step and an uncontrolled formation of foam bubbles is avoided. The pressure p as function of the molar volume V_m of CO_2 ($p(V_m)$-diagram) is shown in Figure 2.14 with the critical point and isotherms below, at and above the critical temperature T_c.

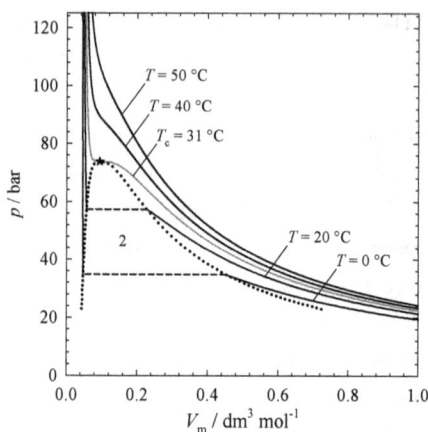

Figure 2.14: $p(V_m)$-diagram, i.e. pressure as function of molar volume, of CO_2. The critical point of CO_2 at $T_c = 30.98$ °C, $p_c = 73.77$ bar and $V_{m,c} = 0.0941$ dm^3 mol^{-1} is marked by an asterisk. Below T_c the isotherms pass (dashed lines) the 2-phase region (dotted line). Expansion above the critical temperature results in a continuous transition from liquid-like to gas-like density, which is proportional to V_m^{-1}. Thus, a nucleation step is avoided. Redrawn from [115], data taken from [116].

Of course, the use of water as the hydrophilic component does not allow for the formulation of stable foams. Instead different materials and a polymerization reaction, of e.g. monomers or polyol, or cooling below the glass transition temperature, of e.g. highly concentrated sugar solutions, are required during or following the expansion to form stable foams. The formulation of efficient and appropriate microemulsions containing the hydrophilic component, the blowing agent and surfactant, and the well-timed fixation are most challenging. This work focuses on the formulation of polyol – CO_2 – surfactant microemulsion systems for the production of polyurethan (PU) foams (see Introduction). Microemulsions of the type highly concentrated sugar solution – supercritical propane – surfactant were already used for the proof of concept [30, 31]. Thereby pore diameters of about 1–30 µm with an additional underlying nanometer-sized substructure of the foam matrix were shown for the sugar foams. While the substructure is the proof of the POSME-concept, µm-scale pores are formed unintentional. These pores can be explained by aging phenomena.

2.1.4 Aging phenomena

Microemulsions with a near or supercritical fluid exhibit two different aging phenomena, namely *Ostwald* ripening [32–34] and coagulation with subsequent coalescence [34–36]. In both cases the driving force is the reduction of the free energy of the system. This is of increasing importance during the expansion step since here the hydrophilic and hydrophobic components come in direct contact.

At high pressures the supercritical fluid, e.g. CO_2, is strongly compressed and dissolved in e.g. H_2O or polyol by the surfactant. Here the surface of the fluid-swollen micelles is completely covered by surfactant molecules, forming the amphiphilic film. Hence, very low interfacial tensions can be found between hydrophilic and hydrophobic component. By decreasing the pressure of the system, the effective volume of the supercritical fluid and consequently the surface of the micelles increase. Thus, the amphiphilic film is stretched and the interface between hydrophilic and hydrophobic component is no longer densely covered by surfactant molecules. Consequently the hydrophilic and hydrophobic components are in direct contact to each other. This results in a strong increase of the interfacial tension, which decreases the thermal stability of the system. This process is shown schematically in Figure 2.15 for $H_2O - CO_2 -$ non-ionic surfactant.

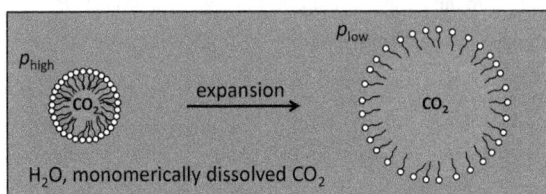

Figure 2.15: Schematic drawing of a CO_2-swollen micelle in H_2O under high pressure (left) and after the expansion, at low pressures (right). At high pressures the interface between H_2O and CO_2 is fully covered by surfactant molecules. By expansion the amphiphilic layer is stretched and H_2O and CO_2 are in direct contact.

As the system with the stretched amphiphilic film and direct contact between the hydrophilic and hydrophobic components is thermodynamically not favored, the system tries to reduce the interface. For a given interfacial tension σ_{ab} a minimization of the interface A_d ($dA_d < 0$) results in a reduction of the free energy F_b of the amphiphilic film [115]

$$dF_b = \sigma_{ab} dA_d \, . \qquad\qquad 2.20$$

Furthermore, the interfacial tension between for instance $H_2O - CO_2$ at $T = 40\,°C$ increases from $\sigma_{ab} = 28$ mN·m^{-1} at $p = 200$ bar to $\sigma_{ab} = 44$ mN·m^{-1} at $p = 50$ bar, i.e. upon expansion [117]. As the determination of interfacial tensions at elevated pressures is sophisticated, extensive studies of the influence of surfactant on the interfacial tension between $H_2O - CO_2$ are missing. However, a value of $\sigma_{ab} \leq 2$ mN·m^{-1} seems to be appropriate [118–121]. In comparison, the interfacial tension of water – oil ($\sigma_{ab} \approx 50$ mN·m^{-1}) can be lowered up to $\sigma_{ab} = 10^{-4}$ mN·m^{-1} by the addition of a surfactant [56].

The minimization of the interface can be achieved by the formation of bigger droplets. According to the surface-to-volume ratio, this means that the surface is more densely packed with surfactant molecules and thus σ_{ab} decreases. However, large droplets are thermodynamically unstable. Thus, the droplets will begin to coagulate until the system is completely phase separated in two coexisting phases. The driving force behind the formation of bigger droplets is the so-called *Laplace* pressure [115]

$$\Delta p_{\text{Laplace}} = \left| p_{\text{inside}} - p_{\text{outside}} \right| = 2\sigma_{ab} H \ .$$
$$\hspace{10cm} 2.21$$

Here p_{inside} and p_{outside} are the pressures inside and outside of the droplet, respectively. Thus, in general, a higher pressure prevails in the inside of a curved surface ($p_{\text{inside}} > p_{\text{outside}}$). Furthermore, $\Delta p_{\text{Laplace}}$ is decreasing with the mean curvature H. For spherical droplets $R_1 = R_2$ (compare Figure 2.9) the *Laplace* pressure can be rewritten as

$$\Delta p_{\text{Laplace}} = \sigma_{ab} \left(\frac{1}{R_1} + \frac{1}{R_2} \right) = \frac{2\sigma_{ab}}{R_0} \ .$$
$$\hspace{10cm} 2.22$$

Consequently, by increasing the mean droplet radius R_0 of the droplets and simultaneously decreasing the interfacial tension, the *Laplace* pressure is strongly reduced. Both possible aging mechanisms for $H_2O - CO_2$ – surfactant systems are shown in Figure 2.16.

The diffusion of CO_2 molecules from smaller to larger droplets forms the basic mechanism of *Ostwald* ripening (Figure 2.16, a)) [33]. According to eq. 2.22 the diffusion of CO_2 from smaller to larger droplets is preferred. Falling below a critical radius, the smaller droplets become unstable and dissolve. The *Ostwald* ripening is particularly relevant for systems exhibiting a high solubility of the fluid in the continuous solvent.

In contrast, the coagulation with subsequent coalescence (Figure 2.16, b)) is diffusion controlled. Temperature induced diffusion of the droplets causes an agglomeration and allow for the formation of one bigger droplet with less stretched amphiphilic film. The diffusion coefficient

D of a particle is related to the viscosity η_0 and the hydrodynamic radius R_h via the *Stokes-Einstein* relation [115]

$$D = \frac{k_B T}{6\pi\eta_0 R_h} .$$ 2.23

t defines the time it takes a particle with the diffusion coefficient D_0 to diffuse a distance x_D. It is defined by the *Einstein-Smoluchowski* equation as

$$t = \frac{x_D^2}{2D} .$$ 2.24

Thus, this process is especially relevant for low viscosity fluids and also dependent upon temperature and the hydrodynamic radius of the droplets.

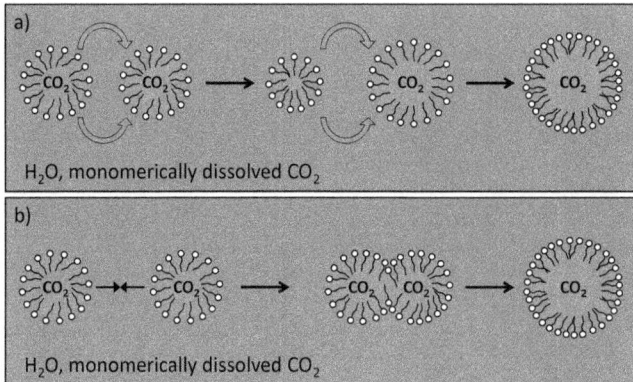

Figure 2.16: Schematic drawing of the two major aging mechanisms of expanded microemulsion droplets. a) *Ostwald* ripening: Diffusion of CO_2 molecules from smaller to bigger micelles until the small micelle dissipates. The higher the monomeric solubility of the fluid in the continuous medium, the faster and more frequent is this process. b) Coagulation with subsequent coalescence: First CO_2-swollen micelles aggregate (coagulation) by thermally induced diffusion. Subsequently a restructuring leads to the formation of one bigger CO_2-swollen micelle with a denser amphiphilic layer. The faster the diffusion of the CO_2-swollen micelles the more frequent is this process.

The PuNaMi-project aims on the production of nanocellular PU foams from polyol-rich CO_2-microemulsion systems. Thereby, pore sizes in the range of or even below 100 nm are envisaged. Thus, aging phenomena are highly undesirable. However, polyols exhibit a relatively large monomeric solubility of CO_2 in polyol of about 10 wt% (see section 4). As a result, *Ostwald* ripening will have a decisive role, especially as long as it is not possible to significantly lower the monomeric solubility of CO_2 in polyol. Furthermore, due to the small droplet size also coagulation with subsequent coalescence is likely. Consequently, one focus of earlier works was the suppression or deceleration of aging phenomena [37–39]. One possible approach is the physical foam stabilizing by spinodal demixing. Thus, the use of a mixture of CO_2 and a low-molecular, hydrophobic co-oil, e.g. cyclohexane, is suggested. Since this co-oil is proposed to avoid aging processes it is called Anti-Aging-Agent (AAA). The composition of the hydrophobic component is described by the parameter

$$\beta = \frac{m_{co\text{-}oil}}{m_{CO_2} + m_{co\text{-}oil}} .$$

2.25

Upon the expansion process the CO_2 – co-oil mixture will undergo a phase separation. According to the binary phase diagram CO_2 – co-oil (Figure 2.17) two different mechanisms for the phase separation can occur. At high pressures the two components CO_2 and AAA are completely miscible, while at low pressures a two-phase region is found due to the formation of gaseous CO_2.

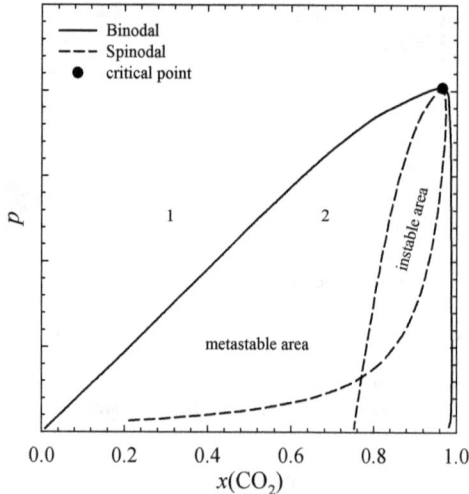

Figure 2.17: Phase behavior of the system CO_2 – low molecular, hydrophobic oil (AAA) as a function of the molar fraction $x(CO_2)$ and pressure at constant temperature. At high pressures a one-phase region is found over the complete x-range. Expansion of the critical composition will result in a spontaneously separation without a nucleation step (spinodal decomposition). At all other molar fractions a metastable area is found below the binodal. Thus, the segregation is much slower. Redrawn from [38].

At most compositions the binodal (also called coexistence curve, solid line) is crossed upon expansion and a metastable area is found. Thus, a nucleation step is required for the phase separation, which implies a comparatively long time scale. Spinodal demixing occurs upon expansion of a sample at or close to the critical composition. Here an instable area is enveloped by the spinodal (dashed line). The phase separation occurs spontaneously (without incubation) by concentration fluctuations, while the amplitude of these fluctuations increases with time [122].

Consequently, mixtures of CO_2 and AAA close to the critical composition are envisaged. It was proposed, that upon spinodal demixing the AAA enriches at the hydrophilic-hydrophobic interface. Thus, a direct contact between the hydrophilic component and the gaseous CO_2 is avoided by the AAA. Furthermore, the interfacial tension σ_{ab} is reduced. The second advantage of using an AAA is the almost pressure-independent density of the co-oil. The proposed formulation of an AAA-enrichment at the water interface and a CO_2-rich droplet core upon expansion is schematically shown in Figure 2.18.

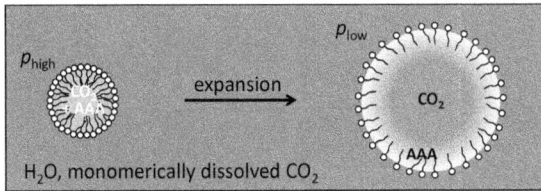

Figure 2.18: Schematic drawing of the expected effect of using a co-oil as Anti-Aging-Agent (AAA) upon expansion. At high pressures (left) the micelles are swollen by an one-phase mixture of CO_2 and AAA. At the critical composition the mixture will decomposes by a spinodal mechanism. In view of the temperature-dependency of the density of CO_2 compared to AAA and the high interfacial tension between CO_2 and the hydrophilic component an enrichment of AAA at the interface is expected.

For the water-rich microemulsion system brine – CO_2/cyclohexane – $F(CF_2)_iC_2H_4E_j$ *Pütz* showed that the addition of an AAA (cyclohexane) indeed successfully decelerates the demixing [59]. By the use of $\beta_i = 0.10$ and $\beta_i = 0.20$ the demixing speed was reduced to 20% and 3% of its initial value, respectively. Thereby the effective CO_2 mass fraction β_i is defined as

$$\beta_i = \frac{m_{co\text{-}oil}}{m_{CO_2,i} + m_{co\text{-}oil}} \qquad\qquad 2.26$$

with $m_{CO_2,i}$ being the mass of CO_2 which is in the hydrophobic phase. Thus, the amount of CO_2 that is monomerically dissolved in the hydrophilic phase was taken into account. It was proposed, that the low monomeric solubility of cyclohexane in brine and thus a reduced *Ostwald* ripening is the main reason for this effect. However, in systematic SANS-measurements it was shown, that even at high pressures a concentration gradient in the micelles instead of a homogeneous distribution of CO_2 and co-oil occurs. Thus, a depletion zone of the co-oil cyclohexane close to the amphiphilic film and a cyclohexane-rich core was found. Figure 2.19 schematically shows the found concentration gradient. The repulsive cyclohexane/fluorinated surfactant tails interaction was postulated to be the reason for this concentration gradient.[59]

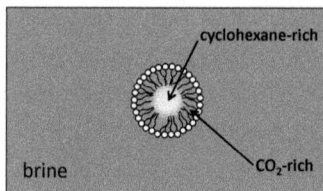

Figure 2.19: Schematic drawing of the concentration gradient of cyclohexane in microemulsions of the type brine – CO_2/cyclohexane – $F(CF_2)_iC_2H_4E_j$ found by *Pütz et al.* [47]. Redrawn from [47].

2.2 Polyurethane

The implementation of the POSME process for the production of nanocellular polyurethane (PU) foams is the main motivation of this work. Therefore this chapter gives a brief introduction to polyurethanes.

The synthesis of urethane from an alcohol and an isocyanate was already described in 1849 by *Wurtz* [3, 123]. However, it was only after the description of the di-isocyanate-polyaddition of linear, branched as well as cross-linked polyurethanes by *Otto Bayer* in 1937, that polyurethanes became ubiquitous with great industrial importance [3, 123–125]. Especially the possibility to tune the properties of polyurethane within a wide range by modifying the raw materials makes them particularly interesting [2, 126]. In general, both foamed as well as solid polyurethane materials are used for a wide range of applications. Foamed polyurethane can be divided into flexible, rigid and integral foam. Flexible PU foams are used for mattresses, upholsteries, in automotive industry, as packing material or thermal insulation of heat accumulators and pipes. As mentioned in the Introduction, rigid PU foams are mainly used for thermal insulation applications, e.g. refrigerators, freezers and insulation boards. Apart from the low thermal conductivities, the self-adhesive effect and the good mechanical properties are exceptional features which characterize rigid PU foams. Integral foams combine a cellular core with a non-cellular skin, which is why they are also called self-skinning foams. Applications for such foams are products in the automotive industry (steering wheels, fenders etc.) and soles of sport and safety shoes. Non-cellular solid polyurethanes are cast or thermoplastic elastomers, coatings, adhesives, fibers or used as binders and wall material of microcapsules. [3]

In general all types of PU are produced from liquid reactive components via an exothermic addition polymerization of an alcohol with an isocyanate, each with a functionality $f \geq 2$ (Figure 2.20).

$$HO-R-OH \quad + \quad OCN-R'-NCO \quad \longrightarrow \quad \text{(polyurethane structure)}$$

Figure 2.20: Generalized addition polymerization of an alcohol with an isocyanate, each with a functionality $f \geq 2$. Redrawn from [3].

A nucleophilic attack of the oxygen of the hydroxyl group at the electrophilic carbon atom of the isocyanate-group forms the new oxygen carbon bond. The reaction is usually accelerated by catalysts. While linear polymer chains are obtained from bifunctional reactants, higher functionalities f are needed for branched or cross-linked PU. In contrast to other polymers, the high reactivity of the polyisocyanates enables a variety of further reactions. Beside the hydroxyl group, amines are the most important reaction partners. The reaction of a primary amine with an isocyanate yields a disubstituted urea (compare Figure 2.23), while trisubstituted ureas are obtained from secondary amines. Generally, isocyanates can react with functional groups containing reactive hydrogen atoms. Isocyanates can thus also react with already formed urethanes and ureas. Furthermore, depending on the catalysts used, isocyanates undergo cycloaddition forming dimers or trimers, reaction to carbodiimides or anionic homopolymerization to 1-nylons.

2.2.1 PU raw materials

The main PU components are polyisocyanates and polyol, while a variety of catalysts and additives are used.

Polyisocyanate

Polyisocyanates can be divided into aromatic and aliphatic ones. Due to higher reactivities towards hydroxyl groups, aromatic polyisocyanates are used for more than 90% of the PU materials. Furthermore, aromatic polyisocyanates improve the mechanical stability. Aliphatic polyisocyanates are mostly used for coating materials to avoid discoloration upon exposure to light and heat. The structural formulas of different, commonly used aromatic and aliphatic polyisocyanates are shown in Figure 2.21.

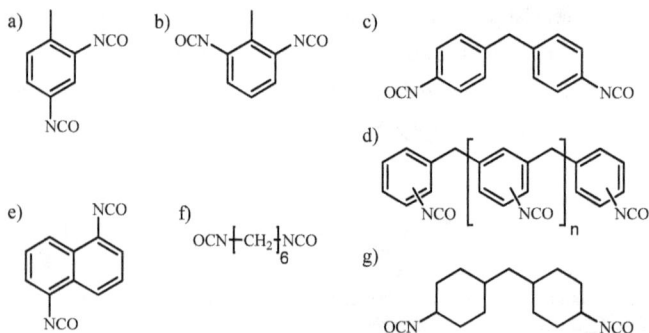

Figure 2.21: Structural formula of different, commonly used polyisocyanates: a) 2,4- and b) 2,6-toluenediisocyanate (2,4-TDI and 2,6-TDI, respectively), c) 4,4'-methylene diphenyl diisocyanate (4,4'-MDI), d) polymeric methylene diphenyl diisocyanate (PMDI), e) 1,5-naphthalene diisocyanate (NDI), f) hexamethylene diisocyanate (HDI), g) 1,1'-methylene bis(4-isocyanatocyclohexane) (H12-MDI). Redrawn from [3].

While different synthetic pathways are known, the industrial, large scale route to polyisocyanates uses the reaction of amines with phosgene [127]. The synthetic pathway to methylene diphenyl diisocyanate (MDI) is shown in Figure 2.22. First nitrobenzene is prepared by nitration of benzene. The nitration takes place by mixing concentrated sulfuric acid and nitric acid [128]. Hydrogenation of nitrobenzene yields aniline, which condensates in the presence of hydrochloric acid with formaldehyde. Subsequent phosgenation of the formed oligomeric di- and polyamines yields polymeric MDI (PMDI), while the isolation of monomeric MDI requires a continuous thin-film distillation. Thereby a mixture of approximately 99% 4,4'-MDI and 1% of 2,4'-MDI is obtained. Excess phosgene is hydrolyzed, while chlorine is recovered from hydrogen chloride by electrolysis. Phosgene is restored by the reaction of chlorine with carbon monoxide in the presence of a catalyst or photochemically.

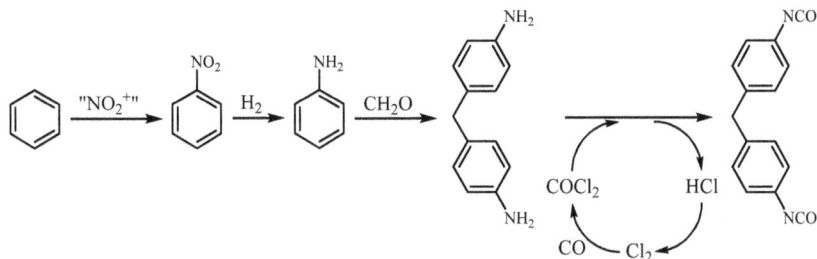

Figure 2.22: Industrial route for the production of methylene diphenyl diisocyanate (MDI). Please note that only the synthesis of the 4,4'-isomer is shown and that a thin-film distillation is required in the last step. Typically first an amine is synthesized, which is treated with phosgene in the following. Thereby phosgene is recovered via hydrogen chloride and chlorine gas by adding carbon monoxide. Redrawn from [127].

Polyols

The use of polyols for the production of PU materials is the main reason for the wide range of properties [3]. In the context of PU the name polyol is used for polymeric organic molecules with in average two or more hydroxyl groups. Thereby the reactivity towards polyisocyanate increases from tertiary via secondary to primary alcohols. Polyether polyols are the most important class of polyols. However, also polyester and polycarbonate polyols are used. Polyether polyols are obtained by alkoxylation of polyfunctional "starter" molecules such as ethylene glycol, glycerol or sorbitol with for instance ethylene oxide (EO) or propylene oxide (PO). Polyester polyols are typically obtained from polycondensation of dicarboxylic acids or their anhydrides with alcohols exhibiting two or more hydroxyl groups. Succinic acid, glutaric acid, phthalic anhydride, ethylene glycol, 1,2-propanediol and glycerol are a small selection of commonly used molecules. Polyester polyols exhibit improved stability towards light and heat [3].

For details concerning the polyols used in this thesis see section 6.2.

Blowing agents

In order to produce a foamed material blowing agents are required [3, 10, 11]. In general, one can distinguish between physical and chemical blowing agents. In case of physical blowing agents a low-boiling liquid or a compressed gas is added to the polyol. Due to the fact, that the polyaddition reaction to urethane is exothermic, the low-boiling liquid evaporates and therewith foams the matrix. In case of a closed-cell structure and blowing agents exhibiting small diffusion coefficients in the PU matrix, the blowing agent remains inside the foam pores. Thus, blowing agents with a low gaseous thermal conductivity $\lambda_{g,0}$ are favorable to reduce the total thermal conductivity λ_{therm}. For instance chlorofluorocarbons (CFCs) were used as physical blowing agents from the late 1950s until the mid-1980s until they were phased out according to the Montreal protocol [129]. Further aspects for selecting the blowing agent are the processability, cost and performance. Blowing agents used are CO_2, fluorinated hydrocarbons (e.g. tetrafluoroethane, pentafluoropropane and pentafluorobutane) or hydrocarbons like n-, iso- and cyclo-pentane, which are used predominantly.

Polyurethane can also be foamed chemically [127]. By the reaction of water and isocyanate carbamic acid is formed as an unstable intermediate (Figure 2.23). Decarboxylation yields an amine and carbon dioxide as the blowing agent, while the amine reacts with another isocyanate forming a disubstituted urea.

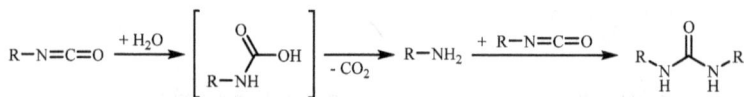

$$R-N=C=O \xrightarrow{+ H_2O} \left[\begin{array}{c} O \\ \parallel \\ R-NH \end{array} OH \right] \xrightarrow{- CO_2} R-NH_2 \xrightarrow{+ R-N=C=O} R \underset{H}{\overset{O}{\underset{N}{\parallel}}} \underset{H}{N} R$$

Figure 2.23: Generalized reaction of isocyanates with water. In the first step carbamic acid is formed, which decomposes spontaneously to an amine and carbon dioxide. The reaction of an amine with an isocyanate yields a disubstituted urea. Redrawn from [3].

At present, in most cases, PU is foamed using a combination of physical and chemical blowing agents. This is because the blowing agent also affects the stability of the foam. Higher boiling blowing agents condense, while CO_2 exhibits a relatively large diffusion coefficient and therefore diffuses out of the foam cell [130]. As a consequence the cell pressure decreases, which results in a shrinkage force. Since these effects mainly occur at low respectively high temperatures, the temperature-dependent properties of the material can be optimized by using a mixture of a physical and a chemical blowing agent.

Additives

Besides blowing agents different additives are used to adjust the properties of the PU material. Catalysts and foam stabilizers are required to control the reactivities and the foaming process, respectively. In general, especially tertiary amines and organotin compounds are used as catalysts for the reaction of isocyanates with alcohols [127]. Foam stabilizers are most often polysiloxane-polyether block copolymers. Thereby their main task is to overcome the destabilizing effect of precipitating urea particles [131]. Furthermore – depending on the PU material as well as the requirements specified – flame retardants are commonly used as additives. Typically, halogenated compounds like tris(2-chloroisopropyl) phosphate (TCPP) are used. In addition, stabilizers towards hydrolysis, oxidation or UV radiation are applied as additives in producing PU materials.

2.2.2 Polyurethane foam processing

Both reactants for the production of PU are typically in liquid form [3]. Thus, an additional solvent is not required in most cases. The industrial processing of PU foams takes place in a one-shot process or a prepolymer process, which involves two- or multiple steps, respectively.

In a one-shot process polyol, polyisocyanate, the blowing agent as well as further additives are mixed directly [3, 132]. Dependent on the catalysts and reactivities the reaction takes 0.5–30 min, while final curing may take further 24–48 h. Essential process parameters are the miscibility, the viscosity and the reactivity of the components. Here, reactivities as uniform as possible are important. The one-shot process predominantly yields statistical products. In contrast tailor-made segment structures are accessible from the controlled polyaddition reaction of prepolymers. Typically, an excess of diisocyanate reacts with polyol in the first step to yield the prepolymer, which is cross-linked in the second step. Thereby the use of prepolymers results in a less exothermic reaction. The prepolymer contains free isocyanate groups or isocyanate-reactive groups. Using isocyanates of different reactivity and thin-film distillation, prepolymers with less than 0.1% monomeric isocyanate are obtained [133, 134]. In addition, a complete reaction of polyols exhibiting low reactivity is accessible without catalysts. However, among other things, due to traces of water and catalyst, these prepolymers are not storage-stable. Furthermore, they are highly viscous, which is disadvantageous with respect to the processability.

A special case for a discontinuous one-shot process is the so-called reaction injection molding (RIM), developed in 1971 [3, 132, 135]. Due to highly reactive components, very short timescales for mixing and injection into the mold are adjusted. Thus the reaction time is in the range of 10–60 seconds. Typical products are automotive components like fenders or spoilers.

In general high- and low-pressure machines are used for the one-shot process (see Figure 2.24) [136]. In both setups two or more working tanks (20–500 liter volume each) are integrated. These tanks are typically equipped with heating/cooling jackets and mechanical agitators. In case of the high-pressure mixing head the components are mixed by injecting them through nozzles with pressures of $p = 100$–250 bar. Thus the kinetic energy and the chaotic flow behavior induced by the impingement causes an efficient mixing. Thereby, a controlled-circulation mixing head of the high-pressure machine allows for a self-cleaning. During the cleaning the raw components are pumped through the mixing head back into the working tanks, while a piston cleans the mixing chamber and the outlet. The high-pressure setup is therefore mainly applied for discontinuous ("one-shot process") productions and outputs of ~ kg/min. The low-pressure setup (Figure 2.24, right), which is less expensive than the high-pressure setup, works at pressures of $p = 3$–40 bar. The raw components are mixed by mechanical agitation while passing the mixing head until the mixture reaches the outlet. Since the mixing head has to be cleaned with a rinsing solution after each shot, the low-pressure setup is mainly applied for the continuous production of foamed PU slabs, blocks or panels.

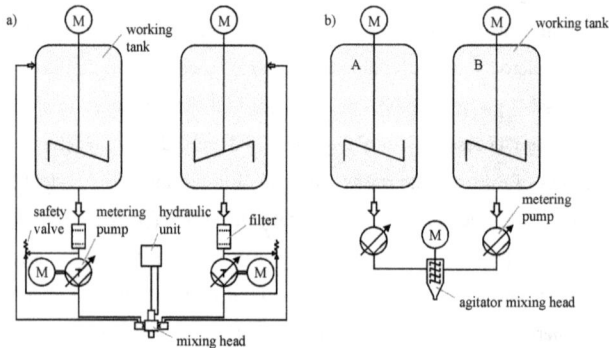

Figure 2.24: Schematic drawing of a) a high-pressure machine with a self-cleaning circulating system and b) a low-pressure PU plant. In high-pressure mixing heads ($p = 100$–250 bar) the component streams are injected via nozzles. Thus the kinetic energy and chaotic flow behavior of the impingement are the driving forces. In the latter case the components are mixed by rotating agitators inside the mixing head at $p = 3$–40 bar. Redrawn from [3].

The insulation of refrigerators etc. implies the foaming of confined but more or less complex cavities. These cavities are filled in a discontinuous process by applying the required amount of reaction mixture in one shot. Thereby one high-pressure mixing head can operate several (movable) molds or only one (stationary) mold. In the case of continuous production the reaction mixture is applied to a conveyor belt. This belt is covered with a paper layer or metal foil, depending on the final product. Additional side and top layers can be used. The conveyor belt allows for a continuous rising and curing of the reaction mixture, while the cured panels or blocks can be trimmed to the desired length.

Furthermore, PU foam can be applied in-situ by spraying or layer-by-layer application. Spray foam is used for insulation as well as sealing purposes of buildings. Thereby the two components pass a heated hose before entering the mixing head. Adjusting the reactivity, curing times within a few seconds are obtained. Cavities of objects that are too large for processing in a factory can be filled by layer-by-layer application. Here a two-component mixture is poured into the cavities.

2.2.3 POSME-pilot plant for the production of PU foams

In 2009 the Bayer MaterialScience AG (BMS) in Leverkusen started up a pilot plant for the technical implementation of POSME. Compared to other commercial PU machines (Figure 2.24) some modifications were made to allow the formulation of foamable CO_2-in-polyol microemulsions. Furthermore, as mentioned in the Introduction, the pilot plant was temporary adapted to a continuous setup after the K 2010 in Düsseldorf. At this time, however, it was not possible to transfer the results from the discontinuous to the continuous setup, which is why again the discontinuous setup was installed. A schematic drawing of the current setup is shown in Figure 2.25. In general, one working tank contains the polyol-surfactant-additive mixture, while a second one contains the polyisocyanate. CO_2, which is provided via a CO_2-tank, and the polyol-surfactant-additive mixture are homogenized by a static mixer, while on the way to the high-pressure mixing head.

Static mixers are in-line mixers using the flow energy for continuous homogenization [137, 138]. They consist of similar, firmly installed mixing units in a tube or channel. E.g. helical elements adjusted to different orientations and angles can be used. Depending on the kind of mixing elements and the flow velocity different lengths are required for homogenization. Furthermore, the laminar or turbulent flow induces a pressure drop. In 2015 BMS realized, that the technical setup does not allow for a complete mixing of CO_2 and the polyol-surfactant-additive mixture. This means that by that time the foaming experiments did not inevitably start from a one-phase state, but probably from a polyol-surfactant-additive mixture more or less saturated with CO_2 and an excess of CO_2. To realize a complete mixing of CO_2 and the polyol-surfactant-additive mixture the length and thus the mixing time of the static mixer was increased. Please note, that the pre-begin of formulating the polyol-CO_2-microemulsions is discarded rather than being fed back into the polyol-surfactant-additive mixture. This would lead to a pre-saturation with CO_2.

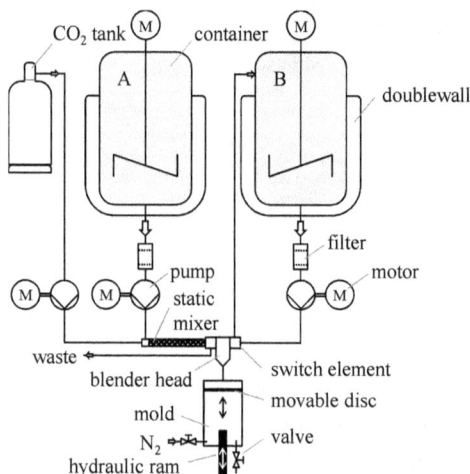

Figure 2.25: Schematic drawing of the discontinuous pilot plant at Covestro. Container A contains the polyol-surfactant-additive mixture, while B contains the isocyanate. A static mixer allows for the homogenization of CO_2 with the polyol-surfactant-additive mixture, while the pre-begin is discarded. To control the position of the movable disc inside the mold pressurized air (counter-pressure) or a hydraulic ram can be used. This allows a controlled foaming process. The pipes as well as the container and the mold are tempered.

The polyol-CO_2-microemulsion formulated in the static mixer is subsequently mixed with polyisocyanate and injected into the mold. A movable disc limits the volume of the mold. Thereby a hydraulic ram or, alternatively, nitrogen counter-pressure allows for a controlled expanding of the reaction mixture. This allows either the system to expand directly or first to harden and then expand.

2.3 Neutron scattering

The phase behavior of microemulsions can be studied by simple experiments. The determination of the structures of microemulsions as well as their sizes and dynamics is however more demanding.[139] Scattering techniques are the most powerful and therefore key techniques to study the structures and dynamics of microemulsions. Direct imaging techniques, like scanning or transmission electron microscopy, exhibit compared to scattering techniques some disadvantages. First, a fixation of the sample, for instance by means of freeze fracture, is required.[54, 90, 97] Thus, local defects may occur and falsify the obtained results. Furthermore, electron microscopy is an invasive method that does not allow for the investigation of the dynamics of systems. Scattering techniques in contrast, are non-invasive and therefore usually do not affect the sample. In addition, they allow for measurements over huge length and time scales and thus enable good statistics. Moreover, dynamic properties and kinetics are accessible via e.g. pressure jump experiments. Scattering experiments can be performed at large-scale facilities or in laboratories. The general setup of scattering experiments is shown in Figure 2.26.

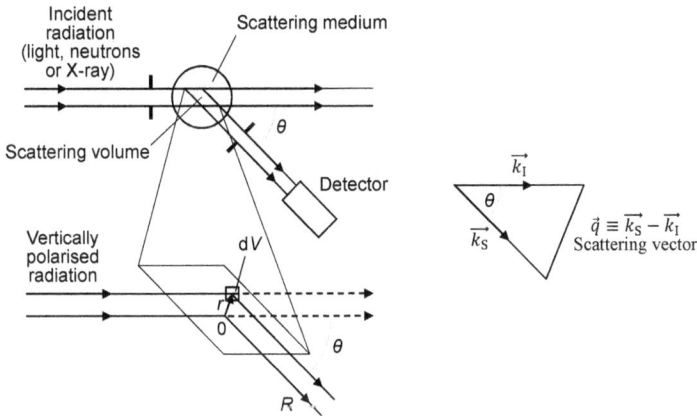

Figure 2.26: Top view of a scattering experiment with an expansion of the scattering volume. The incident radiation (light, neutrons or X-ray) is characterized by the wave vector $\vec{k}_i$ and scattered at the origin O and a volume element dV. The radiation with wave vector $\vec{k}_s$ scattered at the scattering angle θ is analyzed by a detector. The scattering vector $\vec{q}$ is defined as $\vec{q} = \vec{k}_s - \vec{k}_i$. Redrawn from [140].

The incident radiation (light, neutrons or X-ray) with incident wavelength λ_i and energy E_i is delivered by a radiation source and then pass through the sample.[140, 141] According to the wave-particle duality, particles as well as waves can be described by the same principles. The radiation partly passes the sample unaffected, while another part may be absorbed by the sample. Only a small part of radiation interacts with the sample and as a result is scattered (λ_s, E_s). Under a scattering angle θ, a detector measures the scattered radiation. According to the *Fraunhofer* approximation, the wave vectors of the incident $\vec{k_i}$ and the scattered radiation $\vec{k_s}$ can be considered as a planar wave front. This is due to the distances radiation source – sample and sample – detector, respectively, which are much larger than the wavelength of the radiation used.

In general, a scattering process implies an energy transfer. Structural as well as dynamical informations can be obtained from the measured intensity $I = f(\theta, E)$. Assuming "elastic small angle scattering" ($\theta \leq 10°$), energy transfers can be neglected: $\Delta E = 0$ and $\lambda_s = \lambda_i$. Thus, $|k_s| \approx |k_i|$ with the magnitude $2\pi/\lambda$. The so-called scattering vector $\vec{q}$ or "momentum transfer" (Figure 2.26, right) is defined as

$$q = |\vec{q}| = \frac{4\pi}{\lambda}\sin\left(\frac{\theta}{2}\right) .$$ 2.27

However, the scattering is an emission of spherical waves emanating from each scattering center. Consequently, an interference pattern results, which is dependent on the relative positions of the scattering centers to each other. Combining eq. 2.27 with Bragg's law

$$n \cdot \lambda = 2d_s \sin\left(\frac{\theta}{2}\right)$$ 2.28

the scattering vector q is related to the characteristic length scale d_s by

$$q = \frac{2\pi}{d_s}$$ 2.29

with an integer number n.

Beside different radiations, different wavelengths can also be used.[140] While the wavelength of light is ~ 400 nm, the wavelength of neutrons and X-ray is ~ 0.1 nm. As a consequence the scattering vector for neutrons and X-ray is $0.01 \leq q\,/\,\mathrm{nm}^{-1} \leq 10$, which corresponds to length scales of $0.6 \leq d_s\,/\,\mathrm{nm} \leq 600$. The minimum scale for light scattering is ~ 300 nm.[140] In contrast, typical length scales of microemulsions are in the range of $1 - 100$ nm [56]. Thus, especially small angle neutron scattering (SANS) and small angle X-ray scattering (SAXS) are well suited for the investigations of microemulsions. In the case of SANS, the wavelength of the neutron λ is, according to *de Broglie*, defined as $\lambda = h/(m_n v_n)$. With h being the *Planck* constant, m_n the mass of the neutron ($m_n = 1.675 \cdot 10^{-27}$ kg [115]) and v_n their velocity.

However, there is no scattering in homogeneous systems. Scattering of photons is a result of fluctuations in the dielectric properties of the sample, whereas neutrons are scattered by atomic nuclei and X-rays by the electrons of atoms. In case of SANS and SAXS, one can calculate the respective scattering length densities from the individual scattering lengths, which describe the scattering by a single atomic nuclei or electron, respectively.[142] A special feature of neutron scattering lengths is the irregular variation with the type of nuclei. Furthermore, they are sensitive to different isotopes. Hydrogen for instance exhibits a negative coherent scattering length, while the coherent scattering length of deuterium is positive. Consequently, by varying the degree of deuteration, which normally does not affect the systems properties, different contrast conditions are accessible. Thus, it is possible to highlight selected parts of the structure. The X-ray scattering length of atoms in contrast is proportional to the atomic number. By taking into account, that microemulsions mainly consists of carbon, hydrogen and oxygen, contrast variation experiments are therefore more demanding. Therefore, in the following only neutron scattering is considered.

The neutron scattering length density of a domain $\rho_{s,domain}$ consisting of different components is defined as

$$\rho_{s,domain} = \sum_k \phi_k \rho_{s,k} \qquad\qquad 2.30$$

where ϕ_k is the volume fraction of the respective component k in the domain and $\rho_{s,k}$ the respective scattering length density

$$\rho_{s,k} = N_A \frac{\rho_k}{M_k} \sum_i b_i \qquad\qquad 2.31$$

Here, N_A is the Avogadro constant, ρ_k the macroscopic density and M_k the molar mass of the component k and b_i the coherent scattering length of each nuclei i. For a neutron scattering calculator see e.g. [143].

Moreover, X-rays exhibit high initial energies in the order of kiloelectronvolt [144]. The energy resolution is thus not sufficient to study the rather slow motions of soft matter systems. Here energies in the order of $k_B T$ are common. The kinetic energy of thermal neutrons is described via

$$E_n = \frac{h^2}{2m_n\lambda^2}$$

2.32

Thus, at room temperature and for a wavelength of $\lambda = 10$ Å, the energy of a neutron is $E_n \approx 0.03\ k_B T$. Hence, neutron scattering allows for the determination of dynamic processes in soft matters, e.g. microemulsions and polymers.

2.3.1 Small angle neutron scattering

Small angle neutron scattering (SANS) is a static experiment. Thus, an energy transfer to the sample and a change of the wavelength can be neglected. Therefore it is called elastic scattering. The experiment yields informations on the q-dependent scattering intensity $I(q)$ in reciprocal space. As has been noted (see eq. 2.27), the scattering vector q depends on the scattering angle θ and the wavelength λ.

Experimental setup

All SANS-experiments performed during this work were carried out at the Institut Laue-Langevin (ILL) in Grenoble, France. Thereby, most measurements were performed at the D11 instrument, which is the archetype of a long, pinhole geometry spectrometer [145]. The general setup of the D11 instrument is shown in Figure 2.27. The required neutrons are generated from the ILL high flux reactor in a fission chain reaction from highly enriched uranium (^{235}U) [146]. Therefore, a continuous neutron flux of up to $1.5 \cdot 10^{15}$ neutrons per second and cm^2 is produced. Heavy water is used to cool the reactor core and to obtain thermal neutrons with a speed of $v_n \approx 2.2$ km s^{-1}. Liquid deuterium at $T = 25$ K as moderator yields cold neutrons with a speed of $v_n \approx 700$ m s^{-1}. These neutrons with a wavelength of $\lambda = 4.5$–40 Å are provided to the D11 instrument.

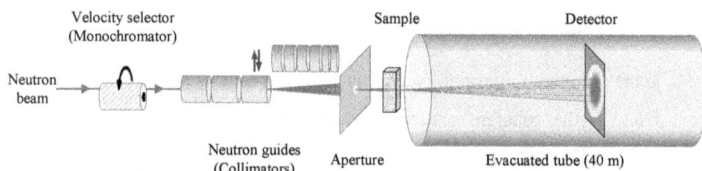

Figure 2.27: Schematic setup of the D11 instrument for small angle neutron scattering. The polychromatic neutron beam is monochromated ($\Delta\lambda/\lambda = 9\%$) using a velocity selector with the wavelength resulting from the rotational speed of the selector. Movable neutron guides allow adjusting the length where the neutron beam is collimated. The collimated and monochromatic beam passes an aperture and a part of the beam is scattered by the sample. The movable detector within the evacuated tube counts the intensity at different pixels. Redrawn from [145].

A mechanical velocity selector with helical slots selects neutrons with a respective wavelength and a distribution of $\Delta\lambda/\lambda = 9\%$ (full width at half maximum), specified by the ILL. The monochromated beam is then collimated by moveable glass guides. The longer the collimation distance is, the lower the beam divergence is, but at the expense of the intensity. Furthermore, the beam can be attenuated by cadmium sheets of different transmissions. After an aperture, the beam passes through the sample and the scattered neutrons are detected via a moveable detector in an evacuated tube. At the D11 spectrometer, a 96 x 96 cm^2 CERCA ^{3}He multi-detector may be placed at twelve discrete sample-to-detector distances of 1.2 to 39 m. Thus, a total scattering vector q of $3 \cdot 10^{-3} \leq q$ / nm^{-1} ≤ 10 is accessible. The detector consists of 128 x 128 pixels with a pixel size of 7.5 x 7.5 mm^2.[145]

All SANS-measurements under elevated pressure were performed using a stroboscopic high pressure SANS-cell with variable volume, which is described in more detail in section 6.3.2. Hellma 404-QX cuvettes (Hellma) and a home-built cell holder with very high temperature stability were used for measurements at ambient pressure (section 6.3.3). For all experiments an aperture with a circular opening of $\emptyset = 13$ mm was used in front of the sample holder. The instrument settings used are listed in more detail in Table 6.8 and Table 6.9.

Data recording and raw data treatment

In general, during SANS-experiments the intensity measured at each detector pixel is detected.[141] As a first step, the beam center and faults have to be masked in the evaluation software. In case of isotropic samples, the intensity is radially averaged, whereby the beam center is obtained from a direct beam measurement with attenuated neutron beam (without sample). The raw data is then corrected for absorption, the sample thickness d_{SA} and the instrument background. Thus, different intensity I_i and transmission T_i calibration measurements are required. Finally the differential scattering cross-section per unit volume of the sample is obtained in the unit of cm^{-1}

$$\left(\frac{d\Sigma}{d\Omega}\right)_{SA}(q) = \frac{I_{SA} - I_{BG} - \dfrac{T_{SA}}{T_{EC}}\left(I_{EC} - I_{BG}\right)}{I_{H_2O} - I_{BG} - \dfrac{T_{H_2O}}{T_{EC\_H_2O}}\left(I_{EC\_H_2O} - I_{BG}\right)} \cdot \frac{T_{H_2O} d_{H_2O}}{T_{SA} d_{SA}} \cdot \left(\frac{d\Sigma}{d\Omega}\right)_{H_2O} \qquad 2.33$$

Here the indices SA, BG, H$_2$O and EC refer to the sample, the background, water and the empty cell. Note, that the SHP-SANS cell and the home-built cell holder for the Hellma 404-QX cuvettes can be corrected using only one water measurement. Therefore, the intensities and transmissions of water and the empty cell, which is used for the water measurement (index EC_H$_2$O), are considered. $(d\Sigma/d\Omega)_{H_2O}$ being the differential scattering cross-section of water. In addition, the thickness of the water sample d_{H_2O} is considered. However, the scattering of water is very weak at sample-to-detector distances larger than 10 m. Thus, the measurement of water at 10 m was used for normalization of the data obtained at larger sample-to-detector distances. Here, the calibration measurement of the H$_2$O-scattering with adequate signal-to-noise ratio would have been too time-consuming. Therefore, a strong scatterer, e.g. Teflon, can be used to determine the scaling factor for the calibration. The standard evaluation software LAMP [147] was used for the raw data treatment at the D11 instrument. Furthermore, the evaluation software considers the dead time of the detector by correcting the transmissions. In addition, the incoherent scattering background of the sample I_{Incoh} must be taken into account. The value of I_{Incoh} depends on the composition, and especially the ratio between hydrogen and deuterium atoms for soft matter systems. The incoherent scattering background was always subtracted from the scattering intensity unless otherwise stated.

In case of measurements at the D22 instrument, the standard evaluation software GRASP [148] was used. Here, instead of water measurements, the incident neutron flux and the incident intensity are used for calibration to absolute scale. For more details concerning the D22 instrument see [149].

In principal there are two options for the analysis of scattering data, which should result in the same information.[150] *Fourier* transformation of the smoothed and desmeared scattering curve yields the pair distance distribution function. From the shape of this function, the geometry and moreover the scattering length density profile can be obtained. However, this is the more demanding way. Alternatively, the experimental data are compared to a scattering curve calculated from a theoretical model. This involves the risk of using too many parameters and an incorrect model. In the following, the different evaluations and the models used are described.

Scattering model for bicontinuous microemulsions

Bicontinuous microemulsions are characterized by two main features.[103, 151] An alternating arrangement of water and oil domains is combined with the absence of long range order. Thus, the microstructure as a function of the distance r can be described by the correlation function $\gamma(r)$

$$\gamma(r) = \frac{d_{TS}}{2\pi r} \cdot \sin\left(\frac{2\pi r}{d_{TS}}\right) \cdot \exp\left(-\frac{r}{\xi_{TS}}\right) \qquad 2.34$$

with the periodicity d_{TS} as a measure of the quasiperodic repeat distance of the water and oil domains and the correlation length ξ_{TS}. The exponential with the correlation length ξ_{TS} represents the absent long range order. *Fourier* transform of eq. 2.34 yields the absolute intensity $I(q)$ according to the *Teubner-Strey* model [103]

$$I(q) = \frac{8\pi c_2 \phi_a \phi_b (\Delta \rho_s)^2 / \xi_{TS}}{a_2 + c_1 q^2 + c_2 q^4}$$ 2.35

With $\Delta \rho_s$ being the scattering length density difference between the two domains a and b, e.g. water and oil, of the respective volume fractions ϕ_a and ϕ_b. The free fit parameters a_2, c_1 and c_2 stem from a *Landau-Ginzburg* order parameter expansion of the local free energy density including gradient terms. The periodicity d_{TS} and correlation length ξ_{TS} can be determined as a combination of the parameters a_2, c_1 and c_2. Thus, d_{TS} and ξ_{TS} are defined by

$$d_{TS} = 2\pi \left[\frac{1}{2} \left(\frac{a_2}{c_2} \right)^{0.5} - \frac{c_1}{4c_2} \right]^{-0.5}$$ 2.36

and

$$\xi_{TS} = \left[\frac{1}{2} \left(\frac{a_2}{c_2} \right)^{0.5} + \frac{c_1}{4c_2} \right]^{-0.5}$$ 2.37

Furthermore, an amphiphilicity factor f_a was described as [152]

$$f_a = \frac{c_1}{(4a_2 c_2)^{0.5}}$$ 2.38

The amphiphilicity factor quantifies the efficiency of the amphiphile in the system under certain conditions (Figure 2.28).[153]

Please note, that the scattering data were described by the expression

$$I(q) = \frac{1}{a_2 + c_1 q^2 + c_2 q^4}$$ 2.39

For strongly structured microemulsions f_a is typically close to -1, while $f_a = -1$ corresponds to the lamellar instability line (see Figure 2.28).[153] The scattering data of strongly structured systems show a pronounced correlation peak ($c_1 < 0$) at $q > 0$. At the *Lifshitz* line ($f_a = 0$), which corresponds to $c_1 = 0$, the correlation peak vanishes. Therefore, the sign of c_1 changes from negative to positive upon crossing the *Lifshitz* line. At the disorder line ($f_a = 1$), the remains of the interfaces become uncorrelated and hence the quasiperiodical order is lost. Furthermore, the periodicity d_{TS} diverges at $f_a = 1$ and the correlation function (eq. 2.34) becomes a simple exponential decay [151, 152]. In line with this, the scattering data decay monotonically with increasing q. At even more positive values of f_a, tricritical points are reached [151, 152, 154]. However, uncorrelated interfaces can still be detected. Note, that crossing the *Lifshitz* line and the disorder line is not associated with macroscopic phase transitions [152].

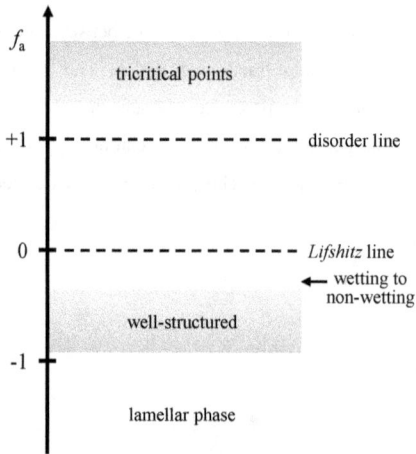

Figure 2.28: Schematic drawing of the amphiphilicity factor f_a, which is independent of the microemulsion system studied. For values of f_a close to -1 (high amphiphilicity) the systems are well-structured. This order is weakened with decreasing amphiphilicity (still $f_a < 0$). By further decrease of the amphiphilicity the *Lifshitz* line ($f_a = 0$) and disorder line ($f_a = 1$) are crossed. At $f_a = 1$, the remains of the interfaces become uncorrelated, i.e. the correlation function becomes an exponential decay. Redrawn from [152].

As pointed out above, the amphiphilic film is fundamental for microemulsions. Theoretical studies showed the high importance of the elastic properties of the amphiphilic film [92, 93, 155–157]. According to *Pieruschka*, *Safran* and *Marcelja* [157], the renormalized bending rigidity κ_{SANS} can be directly calculated from the periodicity d_{TS} and correlation length ξ_{TS} by using the model of random interfaces

$$\frac{\kappa_{SANS}}{k_B T} = \frac{10\sqrt{3}\pi}{64} \frac{\xi_{TS}}{d_{TS}}$$

2.40

While the *Teubner-Strey* model is an appropriate model to describe the scattering data at the scattering peak, it systematic deviates at high q-values.

Porod model

SANS-curves of microemulsions typically exhibit a q^{-4} decay at high q-values, resulting from a sudden change of the scattering length density at the so-called *Porod* limit [158]. Moreover, *Strey, Winkler* and *Magid* assumed a *Gaussian* [159] smoothing function in order to account for the diffusivity of the amphiphilic film. Thus, for bulk contrast ($\rho_{core} = \rho_{film} \neq \rho_{bulk}$), the scattering data at high q-values can be described by

$$\lim_{q \to \infty}[I(q)] = 2\pi \frac{\Delta\rho_s^2}{q^4} \frac{S}{V} \exp(-q^2 t^2)$$ 2.41

with t being the diffusivity parameter. If $t \to 0$ the usual *Porod* limit is obtained. The specific internal interface S/V is defined as

$$\frac{S}{V} = \frac{a_C}{v_C}\phi_{C,i}$$ 2.42

With S/V being the specific internal interface, the surface area a_C and volume v_C per surfactant molecule and the interfacial volume fraction of surfactant $\phi_{C,i}$.

To account for inaccuracies in the absolute calibration of $I(q)$, the specific internal interface can be determined as

$$\frac{S}{V} = \frac{\pi\phi_A\phi_B q^4 \exp(q^2 t^2)}{Q}\left(\lim_{q \to \infty}[I(q)]\right)$$ 2.43

The invariant

$$Q = \int_0^\infty q^2 I(q)dq = 2\pi^2 \langle\eta^2\rangle$$ 2.44

was introduced by *Porod* [158], while $\langle\eta^2\rangle = \phi_a\phi_b \cdot \Delta\rho_s^2$ is the average fluctuation of the scattering length density of the two domains.

The characteristic size of the microstructure of bicontinuous microemulsions ξ is connected to the specific internal interface S/V by

$$\xi = a\frac{\phi(1-\phi)}{S/V}$$ 2.45

with the volume fraction ϕ of oil in the oil/water-mixture and the constant prefactor a ranging from $a = 4$–6 depending on the model used [56, 80, 160, 161].

Debye model

Already in 1949 a special case of eq. 2.35 was described by *Debye* for completely random systems [161, 162]. Thereby, the scattering intensity is defined as

$$I(q) = \frac{8\pi \langle \eta^2 \rangle \xi_D^3}{\left(1 + \xi_D^2 q^2\right)^2} \qquad\qquad 2.46$$

with the characteristic length ξ_D. Note, that eq. 2.46 is the *Fourier* transform of the correlation function $\gamma(r)$

$$\gamma(r) = \exp\left(-\frac{r}{\xi_D}\right) \qquad\qquad 2.47$$

and the scattering intensity consequently monotonic decreases.

Guinier

The *Guinier* analysis is one method to check the experimental scattering data for the existence of spherical, cylindrical or lamellar structures [150, 163]. Therefore, $\ln(I(q) \cdot q^x)$ is plotted as a function of q^2. In the limit of low q-values, a straight line for $x = 0$, 1 or 2 implies the scattering from spherical, cylindrical or lamellar structures, respectively. Thereby this analysis works without further parameters. Moreover, deviations from the straight line at the lower q-limit indicate the occurrence of appreciable particle interference. For spherical particles, the square of the radius of gyration R_g is proportional to the slope of the straight line, while the intensity is [164]

$$I(q) = I(0)\exp\left(-\frac{q^2 R_g^2}{3}\right) . \qquad\qquad 2.48$$

For dilute systems, $I(0)$ is thereby proportional to the volume fraction of the dispersed particles or droplets [165]. Note, that the expression $q_{max} \cdot R_g < 1$ has to be valid. Having checked the experimental scattering data for the existence of spherical, cylindrical or lamellar structures, they can subsequently be compared to a calculated scattering curve for a certain model.

Droplet form factor

According to the decoupling approximation for monodisperse particles the expression

$$\frac{d\sigma(q)}{d\Omega} = n \cdot P(q) \cdot S(q) \tag{2.49}$$

is valid [166]. With n being the number density of scattering particles, $P(q)$ the form factor and $S(q)$ the structure factor. The form factor describes the scattering from a single particle, while the structure factor describes the interference arising from neutrons scattered from different particles. Hence, for very diluted system, i.e. a very low number density of scattering particles, the structure factor becomes $S(q) = 1$ [166]. In general, the form factor is the complex square of the scattering amplitude $A(q)$, which is the *Fourier* transform of the scattering length density distribution function $\Delta\rho_s$ [164]. For monodisperse diluted particles, the amplitude is given by

$$A(q) = \int_V \Delta\rho_s(\vec{r}) \cdot \exp(i\vec{q} \cdot \vec{r}) d\vec{r} \ . \tag{2.50}$$

The differential scattering cross-section for homogeneous particles is thus

$$\frac{d\sigma(q)}{d\Omega} = n \cdot \Delta\rho_s^2 \cdot V_{part}^2 \cdot P(q) \cdot S(q) \tag{2.51}$$

with V_{part} the volume of the scattering particle.

To include the polydispersity of microemulsion droplets, the so-called local monodisperse approximation

$$\frac{d\sigma(q)}{d\Omega} = n \cdot \int_0^\infty dR \cdot S(q, R) \cdot P(q, R) \cdot W_R(R, R_0, \sigma) \tag{2.52}$$

was described [167, 168]. Here the size polydispersity is taken into account by using the *Gaussian* distribution function

$$W_R(R, R_0, \sigma) = \frac{1}{\sqrt{2\pi\sigma^2}} \exp\left\{-\frac{(R - R_0)^2}{2\sigma^2}\right\} \ . \tag{2.53}$$

R being the radius around the mean radius R_0, and σ being the width of the distribution. Consequently, the polydispersity index is given by σ/R_0. Microemulsion droplets typically exhibit polydispersity indices of $\sigma/R_0 \approx 0.25$ [96, 169].

Assuming polydisperse spherical droplets with core scattering, i.e. $\rho_{core} \neq \rho_{bulk} \neq \rho_{film}$, the differential cross-section is given by

$$\frac{d\sigma(q)}{d\Omega}\bigg|_{core} = \phi_{c,1} \cdot \frac{4\pi \cdot a_c}{v_c \cdot q^6 \cdot (R_0^2 + \sigma^2)} \Delta\rho_s^2 \cdot (g_1 + g_2 + g_3) \cdot \exp(-q^2 t^2) \tag{2.54}$$

with

$$g_1 = \frac{1}{2} \left[1 - \cos(2qR_0) \exp(-2q^2\sigma^2) \right]$$

$$g_2 = -q \cdot \left[R_0 \sin(2qR_0) + 2q\sigma^2 \cos(2qR_0) \right] \exp(-2q^2\sigma^2)$$

$$g_3 = \frac{1}{2} q^2 \cdot \begin{bmatrix} -4q\sigma^2 R_0 \sin(2qR_0) \exp(-2q^2\sigma^2) + R_0^2 + \sigma^2 + R_0^2 \cos(2qR_0) \\ \cdot \exp(-2q^2\sigma^2) + \sigma^2 \cos(2qR_0)(1 - 4q^2\sigma^2) \exp(-2q^2\sigma^2) \end{bmatrix} .$$

2.55

Note that, instead of a step function of the scattering length densities, i.e. $\Delta\rho_{core} = \rho_{bulk} - \rho_{core}$ for $0 \leq r \leq R$ and $\Delta\rho_{core} = 0$ elsewhere, a diffuse interface with a standard variation t was considered. The number density of the droplets within the sample is given by the division of the total area of the internal interface $\phi_{C,i} a_C / v_C$ by the interfacial area $4\pi(R_0^2 + \sigma^2)$ [170].

In the case of film contrast conditions, i.e. $\rho_{core} = \rho_{bulk} \neq \rho_{film}$, only scattering from the amphiphilic film is measured. With respect to the diffuse nature of the amphiphilic film, again a *Gaussian* distribution function with the standard deviation t, related to the thickness amphiphilic film, is used. Note, that the differential scattering cross section for film contrast conditions is given by [170]

$$\left. \frac{d\sigma(q)}{d\Omega} \right|_{film} = \phi_{C,i} \cdot \frac{4\pi \cdot v_C}{a_C \cdot q^2 \cdot (R_0^2 + \sigma^2)} \cdot \Delta\rho_{film}^2 \cdot (f_1 + f_2 + f + f_4) \cdot \exp(-q^2 t^2)$$

2.56

with

$$f_1 = \frac{1}{2} q^2 t^4 \left[1 + \cos(2qrR_0) \exp(-2q^2\sigma^2) \right]$$

$$f_2 = qt^2 \cdot \left[R_0 \sin(2qR_0) + 2q\sigma^2 \cos(2qR_0) \right] \exp(-2q^2\sigma^2)$$

$$f_3 = \frac{1}{2} R_0^2 \cdot \left[1 - \cos(2qR_0) \exp(-2q^2\sigma^2) \right]$$

2.57

$$f_4 = \frac{\sigma^2}{2} \left[1 + 4qR_0 \sin(2qR_0) \exp(-2q^2\sigma^2) + \cos(2qR_0)(4q^2\sigma^2 - 1) \exp(-2q^2\sigma^2) \right] .$$

If all three scattering length densities within the investigated sample are unequal, i.e. $\rho_{core} \neq \rho_{bulk} \neq \rho_{film}$, a core-shell structure results. Consequently, the form factor is given by

$$P(q) = \Delta\rho_{film}^2 \cdot V_{film}^2 \cdot A_{film}^2(q) + \Delta\rho_{core}^2 \cdot V_{core}^2 \cdot A_{core}^2(q)$$
$$+ 2\Delta\rho_{film} \cdot V_{film} \cdot \Delta\rho_{core} \cdot V_{core} \cdot A_{film}(q) \cdot A_{core}(q)$$

2.58

with $2A_{\text{film}}(q)A_{\text{core}}(q)$ being the so-called cross term [51]. The differential scattering cross section of the cross term is defined as

$$\left.\frac{d\sigma(q)}{d\Omega}\right|_{\text{cross}} = \phi_{\text{C,i}} \cdot \frac{8\pi}{q^4 \cdot \left(R_0^2 + \sigma^2\right)} \cdot \Delta\rho_{\text{film}} \cdot \Delta\rho_{\text{core}} \cdot \left(h_1 + h_2 + h_3\right) \cdot \exp\left(-q^2 t^2\right) \qquad 2.59$$

with

$$h_1 = \frac{R_0}{2}\left[1 - \cos(2qR_0)\exp\left(-2q^2\sigma^2\right)\right]$$

$$h_2 = q \cdot \sin(2qR_0)\exp\left(-2q^2\sigma^2\right)\left[\sigma^2 + \sigma^2\left(2q^2\sigma^2 - \frac{1}{2}\right) - \frac{R_0^2}{2}\right] \qquad 2.60$$

$$h_3 = q^2\left[\begin{array}{l} t^2 \sin(2qR_0)\exp\left(-2q^2\sigma^2\right)\left(\frac{1}{2q} + q\sigma^2\right) - \frac{R_0 t^2}{2} \\ - R_0 \cos(2qR_0)\left(\frac{t^2}{2} + 2\sigma^2\right)\exp\left(-2q^2\sigma^2\right) \end{array}\right].$$

Again the mean droplet radius R_0, the diffusivity of the amphiphilic film t and the width σ of the distribution of the radius R around the mean droplet radius R_0 are the fit-parameters.

Cylinder form factor for core scattering

Depending on composition, temperature and pressure, microemulsion droplets can elongate to form cylindrical droplets with a length L. For $qR \ll 1$, the form factor $P_{\text{cylinder}}(q)$ can be factorized into an axial contribution of the rod scattering $P_{\text{rod}}(q)$ and a cross sectional contribution $P_{\text{cross}}(q)$. Here, the number density of droplets is given by

$$n = \phi_{\text{C,i}} \frac{a_{\text{C}}}{4\pi v_{\text{C}} R_0(L + R_0)}. \qquad 2.61$$

The differential scattering cross-section is defined as

$$\left.\frac{d\sigma(q)}{d\Omega}\right|_{\text{rod}} = nL^2 \exp(-q^2 t^2) \cdot S(q) \cdot \int_0^\infty dL \cdot P_{\text{rod}}(q) \cdot W_L(L) \cdot \int_0^\infty dR \cdot W_R(R, R_0, \sigma) \cdot P_{\text{cross}}(q) \qquad 2.62$$

with the axial contribution of the rod scattering $P_{\text{rod}}(q)$ and the distribution function W_R (given by eq. 2.53). In addition, an exponential distribution W_L of the cylinder length L with the average length $\langle L \rangle$ has to be considered.[72, 171, 172] W_L is given by [172]

$$W_L(L) = \frac{L}{\langle L \rangle^2} \exp\left(-\frac{L}{\langle L \rangle}\right) .$$

2.63

Under the assumption that the rod is infinitely thin and averaging over all possible orientations in space, the axial contribution can be described as [173]

$$P_{rod}(q) = \frac{2Si(qL)}{qL} - \frac{4\sin^2(0.5 \cdot qL)}{q^2 L^2} .$$

2.64

$Si(qL)$ being the sine integral function of qL

$$Si(x) = \int_0^x \frac{\sin t}{t} dt .$$

2.65

The cross sectional contribution $P_{cross}(q)$ for core scattering can be described as

$$P_{cross}(q) = \Delta\rho_{core,cross}^2 \cdot a_{core,cross}^2 \cdot A_{core,cross}^2(q)$$

2.66

with $a_{core,cross} = 2\pi R^2$ and the amplitude

$$A_{core,cross}(q) = J_1(qR)/qR$$

2.67

with the *Bessel* function J_1.

Cylinder form factor for core and film scattering

Accounting for different contrast conditions and thus different contributions from film and core scattering, the cylinder form factor is given by

$$P_{cylinder}(q) = L^2 \cdot \int_0^\infty dL \cdot P_{rod}(q) \cdot W_L(L) \cdot \int_0^\infty dR \cdot P_{cross}(q) \cdot W_R(R, R_0, \sigma)$$

2.68

with the cross sectional contribution

$$P_{cross}(q) = \left[2\pi \cdot \Delta\rho_{film} \int_0^\infty dr \cdot r \cdot f_{droplet}(r, R) \cdot J_0(qr) \right]^2 .$$

2.69

J_0 is the zeroth order *Bessel* function of first kind, while the radial density distribution function $f_{droplet}(r,R)$ was suggested by *Foster et al.* [174] to

$$f_{droplet}(r, R) = \frac{1}{\exp\left[\dfrac{r - R - d/2}{\chi}\right] + 1} - \frac{1 - \Delta\rho_{core}/\Delta\rho_{film}}{\exp\left[\dfrac{r - R + d/2}{\chi}\right] + 1} .$$

2.70

Here, d is the shell thickness and the parameter χ defines the shape of the scattering length density profile, i.e. the sharpness of the interface.

Structure factor

With respect to the decoupling approximation, the interference of neutrons scattered from different particles is described by the structure factor. A detailed representation of the structure factors is shown in e.g. [22].

The structure factor of monodisperse droplets with repulsive hard-sphere interactions can be analytically solved by using the *Percus-Yevick* approximation for the closure relation. Thus, for a dispersed volume fraction ϕ_{disp} of monodisperse, hard-sphere droplets of diameter d_{HS} the structure factor is given by [175–177]

$$S_{PY}(q) = \frac{1}{1 - n \cdot c(q)} \ .$$

2.71

With $c(q)$ being the *Fourier* transform of the direct correlation function (see [176]). Note that the hard-sphere radius was considered to be $R_{HS} = R_0 + t$, with R_0 being the radius and t the thickness of the amphiphilic film [168]. $n \cdot c(q)$ can be calculated to

$$n \cdot c(q) = \frac{1}{x^3} \cdot \left\{ \begin{array}{l} A \cdot (\sin x - x \cdot \cos x) + B \cdot \left[\left(\frac{2}{x^2} - 1 \right) \cdot x \cdot \cos x \right] \\[2mm] + B \cdot \left(2 \cdot \sin x - \frac{2}{x} \right) \\[2mm] - \frac{A \cdot \phi_{disp}}{2} \cdot \left[\frac{24}{x^3} + 4 \cdot \left(1 - \frac{6}{x^2} \right) \cdot \sin x - \left(1 - \frac{12}{x^2} + \frac{24}{x^4} \right) \cdot x \cdot \cos x \right] \end{array} \right\}$$

2.72

with $x = q \cdot 2R_{HS}$ and the constants

$$A = -24 \cdot \phi_{disp} \cdot \frac{(1 + 2\phi_{disp})^2}{(1 - \phi_{disp})^4}$$

2.73

and

$$B = 36 \cdot \phi_{disp}^2 \cdot \frac{(2 + \phi_{disp})^2}{(1 - \phi_{disp})^4} \ .$$

2.74

However, equation 2.72 overestimates the oscillations caused by the interparticle interactions [72, 170]. To consider the polydispersity of the microemulsion droplets, the averaged *Percus-Yevick* structure factor [22]

$$\overline{S_{PY}(q)} = \int_0^\infty dR_{HS} \cdot S_{PY}(q) \cdot W_{HS}\left(R_{HS}, R_{HS,0}, \sigma_{HS}\right) \qquad 2.75$$

with the *Gaussian* distribution function $W_{HS}(R_{HS}, R_{HS,0}, \sigma_{HS})$

$$W_{HS}\left(R_{HS}, R_{HS,0}, \sigma_{HS}\right) = \frac{1}{\sqrt{2\pi\sigma_{HS}^2}} \exp\left\{ -\frac{\left(R_{HS} - R_{HS,0}\right)^2}{2\sigma_{HS}^2} \right\} \qquad 2.76$$

has to be applied. σ_{HS} is the deviation of the hard-sphere radius R_{HS} around the mean hard-sphere radius $R_{HS,0}$. Note, that this equation has to be solved numerically.

While the *Percus-Yevick* structure factor $S_{PY}(q)$ helps describing the scattering intensity at intermediate to low q-values, the very low q-region can be described by the *Ornstein-Zernike* structure factor $S_{OZ}(q)$. Thus, an increase of the scattering intensity caused by critical fluctuations is taken into account. Thereby the *Ornstein-Zernike* structure factor is defined as [166, 178, 179]

$$S_{OZ}(q) = 1 + \frac{S_{OZ}(0)}{1 + q^2 \xi_{OZ}^2} \qquad 2.77$$

with $S_{OZ}(0) = n \cdot k_B T \cdot \chi_T$. ξ_{OZ} being the correlation length of critical fluctuations and χ_T being the isothermal compressibility. Eq. 2.77 is the *Fourier* transform of the correlation function

$$\gamma(r) \sim \frac{1}{r} \exp\left(-\frac{r}{\xi}\right). \qquad 2.78$$

2.3.2 Neutron spin echo

Inelastic neutron scattering, e.g. by using neutron spin echo (NSE), is a powerful tool to study dynamic processes [180, 181]. All NSE-measurements were carried out at the IN15 instrument at the ILL in Grenoble using the stroboscopic high pressure SANS-cell (see section 6.3.4). This high energy and momentum resolution spin-echo spectrometer enables distances of 1 to 500 Å and relaxation times of 0.001 to 1000 ns [182]. In general, all inelastic neutron scattering techniques measure the momentum $\vec{q} \equiv \vec{k}_s - \vec{k}_i$ and energy transfer $\hbar\omega \equiv E_s - E_i$ ($\hbar$ being the reduced Planck constant) [183]. Thereby, NSE determines the velocity of the initial and scattered neutrons using the Larmor precession of the neutron spin in a magnetic field. Consequently, the velocity and thus energy change of each neutron upon interaction with the sample is determined. Since the energy resolution is decoupled from the monochromatization, rather large wavelength distributions can be used. The ILL specifies $\Delta\lambda/\lambda = 15\%$ (full width at half maximum) for the

IN15. The schematic setup of a neutron spin echo spectrometer is shown in Figure 2.29. The first "π/2 flipper" flips the spin polarization perpendicular to the main magnetic field. Thus, the Larmor precession of the incident polarized neutrons is initiated. The neutrons are scattered by the sample and in case of inelastic scattering the neutron spin is more or less altered.

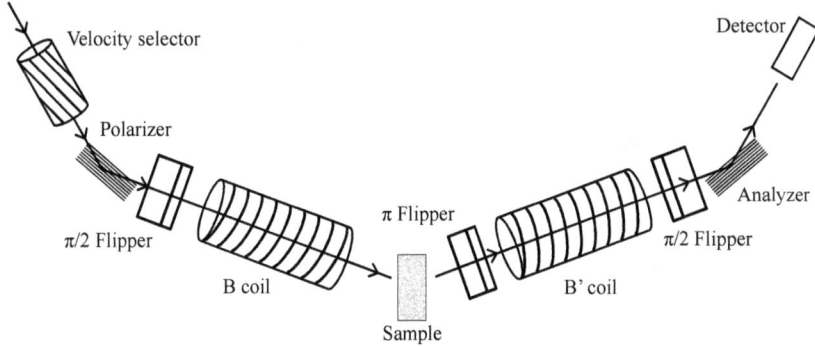

Figure 2.29: Schematic setup of a neutron spin echo (NSE) spectrometer. A wavelength distribution ($\Delta\lambda/\lambda$) of about 15% of the primary neutron beam is selected via a velocity selector. The Larmor precession is initiated by passing the first "π/2 flipper", while the beam was polarized in forward direction by a supermirror. The B and B' coil generate precessions fields parallel to the primary and secondary flight path, respectively. The neutrons are scattered at the sample, whereby in case of inelastic samples the spin is more or less altered. A "π flipper" close to the sample changes the precession angle from α_n to $-\alpha_n$. The precession is stopped by the second "π/2 flipper". The difference of precession angle can be detected and evaluated. Redrawn from [180].

The π flipper close to the sample position reverses the precession angle from α_n to $-\alpha_n$. The second "π/2 flipper" flips the spins along the flight direction and thus stops the Larmor precession. In case of IN15, a two-dimensional 32x32 detector (each 1 cm^2) is positioned at a sample-to-detector distance of 4.6 m [182].

In case of elastic scattering the incident polarization is restored, since the magnetic fields B and B' are equal. For inelastic samples, the difference of the precession angles in $B \cdot l$ and $B' \cdot l'$ is a measure of the inelasticity of the sample, with the coil length l and l'. Considering $B \cdot l = B' \cdot l'$ and inelastic scattering processes with small energy transfers, the difference of the precession angle $\alpha_n' - \alpha_n$ can be approximated by [180]

$$\alpha_n' - \alpha_n = \left(\frac{2\pi |g_n| \mu_n m_n}{h^2} \right) Bl(\lambda_s - \lambda_i) \approx \frac{|g_n| \mu_n m_n^2 \lambda^3 Bl}{h^3} \omega = \tau(B)\omega \qquad 2.79$$

with $\Delta\lambda \approx \hbar\omega / \dfrac{dE}{d\lambda}$.

Here, g_n is the gyromagnetic ratio (more precisely the dimensionless magnetic moment) of the neutron ($g_n = -3.83$), μ the nuclear magneton, h the Planck constant and τ the *Fourier* time. Note that during the NSE experiment, τ is varied by adjusting different magnetic fields B. The factor $\cos(\alpha_n{'}-\alpha_n)$ is averaged and weighted by $S(q,\omega)$ to obtain the effective polarization $P(q,\tau)$, which is defined as

$$P(q,\tau) = \frac{\int\limits_{-\infty}^{\infty} S(q,\omega)\cos(\omega\tau)d\omega}{\int\limits_{-\infty}^{\infty} S(q,\omega)d\omega} .$$

2.80

Fourier transform of $S(q,\omega)$ yields the intermediate scattering function $S(q,\tau)$

$$S(q,\tau) = \int\limits_{-\infty}^{\infty} S(q,\omega)\exp(i\omega\tau)d\omega = \int d^3r \exp(i(q\cdot r))G(r,\tau) .$$

2.81

The normalized intermediate scattering function

$$P(q,\tau(B)) = \frac{S(q,\tau(B))}{S(q,0)}$$

2.82

directly results from the NSE experiment. For more details concerning the theory and instrumental techniques of NSE see e.g. [181, 183].

Dynamics of bicontinuous microemulsion systems

At high q-values, i.e. in the q-region of the *Porod* model, thermally induced fluctuations so-called undulations of the amphiphilic film of bicontinuous microemulsions are investigated.[184] As the membrane interactions are on a much shorter length scale than the membrane-membrane distances, the static and dynamic behavior of the amphiphilic film can be studied independently of possible long range order. Thereby a stretched exponential decay of the dynamic structure factor $S(q,\tau)$ with an exponent ~ 0.7 and relaxation rate proportional to $\kappa^{-1/2}$ was found by *Zilman* and *Granek*. With increasing bending rigidity κ, i.e. stiffness of the amphiphilic film, increasing undulation relaxation rates and decreasing average undulation amplitudes were found. Additional contributions from translational diffusion can be considered by multiplying with the exponential function $\exp(-Dq^2\tau)$. Here D is the diffusion coefficient $D \sim k_B T/\eta_0\xi$, with η_0 being the viscosity and ξ the correlation length. Translational diffusion is expected to be negligible for $q\xi \gg 1$.[184]

According to *Zilman* and *Granek*, for $\kappa/k_B T \geq 1$ and large q, the dynamic structure factor can be represented by the stretched exponential

$$S(q,\tau)/S(q,0) = \exp\left(-\left(\Gamma_{ZG}(q)\cdot\tau\right)^{\beta_{ZG}}\right) \qquad 2.83$$

with the exponent β_{ZG}, which was determined to be close to 2/3 [41, 184, 185]. Here the relaxation rate Γ_{ZG} is given by

$$\Gamma_{ZG}(q) = 0.025\gamma_k \left(\frac{k_B T}{\kappa}\right)^{1/2} \frac{k_B T}{\eta_0} q^3 \qquad 2.84$$

with

$$\gamma_k = 1 - 3\left(\frac{k_B T}{4\pi\kappa}\right)\ln(q\xi_{TS}). \qquad 2.85$$

ξ_{TS} is the correlation length obtained from the *Teubner-Strey* model.

However, a quantitative determination of the bending rigidity according to the *Zilman-Granek* model requires numerical integrations.

In general, the dynamic structure factor can be expressed as

$$S(q,\tau) \propto \left\langle\left\langle\int d^2r\int d^2r' \exp\left(i\vec{q}_\parallel(\vec{r}-\vec{r}')\right)\cdot \exp\{iq_z[h(\vec{r},\tau)-h(\vec{r}',0)]\}\right\rangle\right\rangle \qquad 2.86$$

Where $\vec{q}_\parallel$ is the two-dimensional component of q parallel to the membrane average plane, $q_z = (k_B T/\kappa)^{1/2}\vec{q}_\parallel$ and $h(\vec{r},\tau)$ the local membrane displacement from an average reference plane. Solving of the q_z-term by means of Cumulant expansion yields

$$S(q,\tau) \propto \left\langle\int d^2r\int d^2r' \exp\left(i\vec{q}_\parallel(\vec{r}-\vec{r}')\right)\cdot \exp\left(-q_z^2\left\langle[h(\vec{r},\tau)-h(\vec{r}',0)]^2\right\rangle\right)\right\rangle. \qquad 2.87$$

This becomes

$$S(q,\tau) \propto \left\langle\int d^2r\int d^2r' \exp\left(i\vec{q}_\parallel(\vec{r}-\vec{r}')\right)\right.$$

$$\left.\cdot\exp\left(-\frac{1}{4\pi^2}\cdot\frac{k_B T}{\kappa}\cdot q_z^2\cdot\int_{k_{min}}^{k_{max}} dk\frac{1}{k^4}\left[1-\exp\left(i\vec{k}(\vec{r}-\vec{r}')-\omega(k)\tau\right)\right]\right)\right\rangle_\alpha \qquad 2.88$$

while k labels the undulation mode, the so-called undulation wave vector [185]. In general, the expression

$$\int d^2r \exp(i\vec{k}\vec{r}) = \int\limits_0^\infty \int\limits_0^{2\pi} d\phi dr r \exp(ikr \cos(\phi)) = 2\pi \int dr r J_0(kr) \qquad 2.89$$

is valid, while J_0 being the *Bessel* function of order 0 [185]. Note that the membrane patches are randomly orientated with respect to q. Thereby the angles α between the normal to the patch and q are isotropic between 0 and π. The distribution $g(\alpha)$ is defined as [185]

$$g(\alpha) = \frac{d\Omega}{4\pi} = \frac{2\pi \sin(\alpha)d\alpha}{4\pi} = \frac{1}{2} d\cos(\alpha) . \qquad 2.90$$

The combination of eqs. 2.88, 2.89 and 2.90 yields

$$S(q,\tau) = \frac{2\pi\xi^2}{a^4} \int\limits_0^1 d\mu \int\limits_0^{r_{max}} dr\, r\, J_0\left(qr\sqrt{1-\mu^2}\right)$$
$$\cdot \exp\left(-\frac{k_B T}{2\pi\kappa} \cdot q^2\mu^2 \cdot \int\limits_{k_{min}}^{k_{max}} dk\, \frac{1-J_0(kr)\cdot\exp(-\omega(k)\cdot\tau)}{k^3}\right) \qquad 2.91$$

with μ being the cosine of the precession angle α. The real space upper cutoff is set to $r_{max} \approx \pi/k_{min} = \xi_{TS}/\varepsilon_{ZG}$.[184, 185] Note that the parameter ε_{ZG} is expected to be $\varepsilon_{ZG} \approx 1$. In order to account for the finite size of the surfactant molecules, the upper cutoff of the undulation wave vector was set to $k_{max} = \pi/l_C$ with l_C being the length of a surfactant molecule.[185]

The viscous flow of the embedding membrane induces friction as counteract to the membrane bending. Thus, the *membrane Zimm dynamics* results to the dispersion relation for membrane undulations [186, 187]

$$\omega(k) = \frac{\kappa}{4\eta_0} k^3 . \qquad 2.92$$

Note that the k^3-dependency results from the k^4 scaling of the *Helfrich* free energy and the k^{-1}-dependency of the long range hydrodynamic interactions.[184]

Thus the only fit parameter in eq. 2.91 is the bending rigidity κ. Due to the integration over the undulation wave vector k, the bending rigidity is free from contributions by undulation. Consequently, the bending rigidity as determined from NSE via numerical integrations of eq. 2.91 equals the renormalization corrected bending rigidity $\kappa_{NSE} = \kappa_0$.[41]

3 Aqueous carbon dioxide microemulsions

The study of microemulsion systems of the type water $-CO_2-F(CF_2)_iC_2H_4E_j$ is mainly motivated by the fact, that although the properties of these systems have been studied over the last years in detail, it hadn't been proven to what extent the corresponding state description for water $-n$-alkane $-C_iE_j$ microemulsion systems can be used to describe and predict the properties of CO_2-microemulsions. In order to proof the applicability, the phase behavior of the water-rich, balanced (equal amounts of water and CO_2) as well as CO_2-rich system was studied (see section 3.1.1). Performing SANS-measurements, the transition from CO_2-swollen micelles in water via bicontinuous structures to water swollen micelles in CO_2 was revealed for the first time (section 3.1.2). The similar phase behavior of the temperature dependence of the characteristic length scale found for CO_2- and classical water/oil-microemulsions allowed to scale the $\xi(T)$-data onto each other by plotting the reduced characteristic length scale $\xi \cdot \phi_{C,i,m}$ versus the reduced temperature $2(T-T_m)/T_u-T_l$ (see section 3.1.3). Furthermore, SANS and neutron spin echo (NSE) measurements were used to determine the bending elastic properties of the amphiphilic film (section 3.1.4).

Based on the studies by *Pütz et al.* [47], who found an efficiency boosting, i.e. a reduction of the amount of surfactant needed to formulate a one-phase microemulsion by a factor of 2 to 5, upon partial replacement of CO_2 by the co-oil cyclohexane, NSE-measurements were conducted to study the influence of cyclohexane on the bending elastic properties of the amphiphilic film. In the first step, the phase behavior of the balanced microemulsion system water $-$ CO_2/cyclohexane $-F(CF_2)_iC_2H_4E_j$ was studied, while *Pütz et al.* [47] have studied the water-rich system. The microstructure and bending elastic properties were then studied by means of SANS and NSE.

3.1 Proof of the corresponding state description

In the nineties, *Strey* and *Sottmann* could show that similar to the corresponding state description of real gases by *van der Waals* in 1881 [52] a corresponding state description should exist for microemulsion systems of the type water $-n$-alkane $-C_iE_j$ [53]. Thus, the temperature dependence of the phase behavior, oil/water-interfacial tension σ_{ab} and characteristic length scale ξ of these systems could be scaled onto each other [53–56]. However, until now it was not proven, whether a corresponding state description also exists for water $- CO_2 - F(CF_2)_iC_2H_4E_j$ microemulsion systems. In recent years, several studies on the phase behavior, microstructure and kinetics of water $- CO_2 -$ non-ionic surfactant microemulsion systems were performed [41, 45–51]. Thereby, partly fluorinated, environmental unfriendly surfactants turned out to solubilize water and CO_2 efficiently [42, 43]. *Klostermann et al.* performed systematic studies on the water-rich as well as on the balanced $H_2O/NaCl - CO_2 - F(CF_2)_iC_2H_4E_j$ microemulsion systems [30, 41, 45]. In their study they used the technical polyfluoroalkyl-polyglycolether surfactants Zonyl® FSO 100 (in average $F(CF_2)_8C_2H_4E_7$) and Zonyl® FSN 100 (in average $F(CF_2)_8C_2H_4E_{10}$) produced by DuPont [30, 44, 109, 188]. These surfactants can be synthesized from fluorotelomer alcohols $F(CF_2)_iC_2H_4OH$ (*i* is even) via ethoxylation, while degradation products are e.g. perfluorooctanoic acid (PFOA) or PFOA-related substances [189]. PFOA is known to be a persistent, bioaccumulative and toxic substance [189]. Thus, already in 2006 the U.S. EPA 2010/15 PFOA Stewardship Program was founded to reduce the release of PFOA into the environment [190]. DuPont and seven further major manufacturers of fluoropolymers and fluorotelomers committed to a successive reduction of the production of PFOA or PFOA-related substances. As a consequence the production of the Zonyl® surfactants was phased out. The Capstone® fluorosurfactants were introduced as sustainable replacements. Based on short-chain molecules, they decompose to shorter chain perfluorocarboxylic acids, which are known to be better biodegradable. Thereby Capstone® FS-3100 is suggested to be a suitable replacement for the used Zonyl® FSO 100 and FSN 100 [191]. Hence, with respect to the formulation of a more environmental friendly water $- CO_2 -$ Capstone® FS-3100 microemulsion, the study of the phase behavior is the first step. The experimental methods are described in the appendix (section 6.3).

3.1.1 Phase behavior

In order to validate that the properties of CO_2-microemulsions can be scaled on the ones of classical water/oil-microemulsions, the phase inversion of a CO_2-microemulsion, i.e. the phase behavior of the water-rich, balanced and also CO_2-rich microemulsion system of the type $H_2O/salt - CO_2 -$ Capstone® FS-3100 was studied in a first step. A small amount of salt ($\varepsilon = m_{salt}/(m_{salt}+m_{H_2O})$) was added to suppress possible electrostatic interactions, induced by impurities of the technical grade surfactant.

In a first set of experiments, the phase behavior of the balanced microemulsion system $H_2O/NaCl - CO_2 -$ Capstone® FS-3100 was studied using NaCl as salt with $\varepsilon = 0.01$ and a constant mass fraction $\alpha = 0.40$, i.e. CO_2 – water ratio. The phase behavior was then studied as a function of temperature T and surfactant mass fraction γ. The exemplary $T(\gamma)$-phase diagram at $p = 200$ bar is shown in Figure 3.1. For reasons of better comparison, the respective phase behavior of the microemulsion system $H_2O/NaCl - CO_2 - Zonyl®$ FSO 100/FSN 100 ($\delta_{FSN} = m_{FSN}/(m_{FSO}+m_{FSN}) = 0.75$) is added, which was determined by *Klostermann et al.* [41].

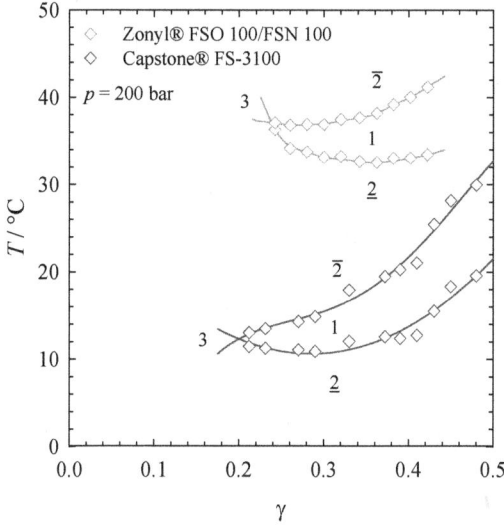

Figure 3.1: $T(\gamma)$-phase diagrams of the system $H_2O/NaCl - CO_2 - F(CF_2)_iC_2H_4E_j$ at $\alpha = 0.40$, $\varepsilon = 0.01$ and $p = 200$ bar for Capstone® FS-3100 (blue diamonds) and Zonyl® FSO 100/FSN 100 ($\delta_{FSN} = 0.75$, grey diamonds). The Zonyl®-containing system was determined by *Klostermann et al.* [41].

Both systems exhibit the phase sequence $\underline{2} \rightarrow 1 \rightarrow \overline{2}$ with increasing temperature, which is typical for microemulsions stabilized by a non-ionic amphiphile. However, the one-phase region of the Capstone®-stabilized system is located at significantly lower temperatures than the Zonyl®-stabilized system. Comparing the $\tilde{X}$-points it turned out, that the replacement of the Zonyl®-surfactants by Capstone® FS-3100 shifts the $\tilde{X}$-point by $\Delta\tilde{T} \approx 24$ °C to lower temperatures, of around $\tilde{T} \approx 12$ °C. Moreover, the surfactant mass fraction $\tilde{\gamma}$ at the optimum point $\tilde{X}$ decreases from $\tilde{\gamma} \approx 0.24$ to $\tilde{\gamma} \approx 0.20$. Note, that due to the lower temperatures the CO_2 is not a supercritical but a near-critical liquid. The shift to lower temperatures indicates, that

Capstone® FS-3100 is more hydrophobic than the surfactant mixture Zonyl® FSO 100/Zonyl® FSN 100 with $\delta_{FSN} = 0.75$.

In order to shift the one-phase region to higher temperatures sodium chloride (NaCl) was exchanged by sodium perchlorate ($NaClO_4$). According to the Hofmeister series NaCl is a lyotropic salt while $NaClO_4$ is a hydrotropic one [192, 193]. Lyotropic salts render non-ionic amphiphiles effectively more hydrophobic, while hydrotropic ones make them more hydrophilic. Thus, the upper two-phase region in the binary water – non-ionic surfactant (A – C) side-system shrinks upon addition of a lyotropic salt. Consequently, the phase boundaries of the ternary microemulsion system water – oil – non-ionic surfactant are shifted to higher temperatures. In view of determining the phase behavior of the water-rich, balanced and CO_2-rich microemulsion, NaCl was thus replaced by $NaClO_4$ adjusting $\varepsilon = 0.05$.

i)   Water-rich microemulsions

The phase behavior of the $T(w_B)$-section of the water-rich microemulsion system $H_2O/NaClO_4 - CO_2 -$ Capstone® FS-3100 at $\gamma_a = 0.08$, $\varepsilon = 0.05$ and 4 pressures between $p = 150$ and 300 bar is shown in Figure 3.2.

As described in section 2.1.1, in general the phase sequence $\underline{2} \rightarrow 1 \rightarrow \overline{2}$ was observed, which is typical for microemulsions stabilized by a non-ionic surfactant [71–73]. At high temperatures a water-in-CO_2 (w/c) microemulsion coexists with a water-excess phase ($\overline{2}$), while at low temperatures a CO_2-in-water (c/w) microemulsion phase coexists with a CO_2-excess phase ($\underline{2}$). By increasing the overall oil mass fraction w_B the upper *near critical boundary* (*ncb*) descends monotonically, i.e. no minimum in the upper phase boundary was observed. A minimum is observed if an additional closed loop is formed in the phase prism, which is typically found for long chain surfactants. Thereby two phases, a dense and a diluted cylindrical network phase, coexist. However, a high salt concentration of 5 wt% is used, while the stronger electrostatics might cause the absence of the closed loop and thus the absence of the minimum in the *ncb*. At $p = 150$ and 200 bar the lower phase boundary $\underline{2} \rightarrow 1$, the so-called CO_2-*emulsification failure boundary* (*cefb*), intersects with the *ncb* at $w_B \approx 0.121$ and the lower critical end point temperature $T_l \approx 10.7$ °C. For $p = 250$ and 300 bar no *cefb* was detected in the investigated temperature range. As can be seen in Figure 3.2, except for the *ncb* at $p = 150$ bar, the pressure only slightly influences the phase behavior of the system. This finding can probably be attributed to the density of CO_2. The density of supercritical CO_2 is strongly pressure and temperature dependent. However, at temperatures of $T < T_c$ the density of liquid CO_2 is only slightly pressure and temperature dependent (ρ_{CO_2}(150 bar, 10 °C) = 0.95 g·cm^{-3}, ρ_{CO_2}(300 bar, 10 °C) = 1.02 g·cm^{-3}) [29]. Note, that a higher CO_2-density results in better CO_2-surfactant interactions and thus better solubilisation of the polyfluorinated surfactant tails by CO_2.

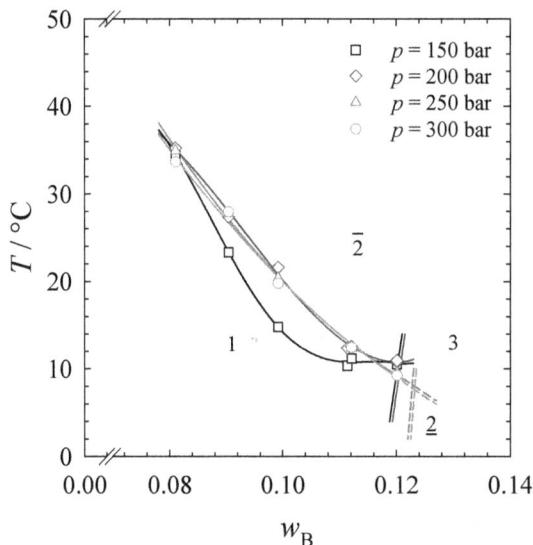

Figure 3.2: $T(w_B)$-phase diagrams of the water-rich microemulsion system $H_2O/NaClO_4$ – CO_2 – Capstone® FS-3100 at $\gamma_a = 0.08$, $\varepsilon = 0.05$ and 4 pressures between $p = 150$ and 300 bar. Upon pressure increase the efficiency of the surfactant to solubilize CO_2 in $H_2O/NaClO_4$ increases only slightly. This might be due to the low temperatures and thus small differences in the density of CO_2 between $p = 150$ and 300 bar. Please note, that in the investigated temperature range no lower phase boundary was detected at $p = 250$ and 300 bar.

ii) Balanced microemulsions

In the next step the amount of CO_2 was increased to study the balanced CO_2-microemulsion system. The phase behavior of balanced microemulsions is commonly studied with $T(\gamma)$-sections by performing vertical sections through the phase prism at a constant mass fraction α. The phase behavior is then determined as a function of temperature T and the overall surfactant mass fraction γ adjusting different pressures. The recorded $T(\gamma)$-sections through the phase prism of the balanced microemulsion system $H_2O/NaClO_4 – CO_2 –$ Capstone® FS-3100 ($\alpha = 0.40$, $\varepsilon = 0.05$) are shown in Figure 3.3.

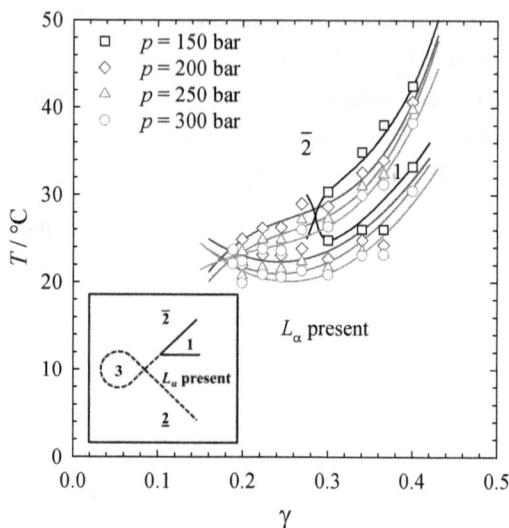

Figure 3.3: $T(\gamma)$-phase diagrams of the system $H_2O/NaClO_4 - CO_2$ – Capstone® FS-3100 at $\alpha = 0.40$, $\varepsilon = 0.05$ and pressures of $p = 150$, 200, 250 and 300 bar. With increasing pressure all phase boundaries are slightly shifted to lower temperatures. Furthermore, especially between $p = 150$ bar and $p = 200$ bar, the efficiency of Capstone® FS-3100 to solubilize $H_2O/NaClO_4$ and CO_2 strongly increases with increasing pressure. Note that at low temperatures a microemulsion phase coexists with a lamellar phase (L_α present) was found (see Figure 3.8). A probable scheme of the phase behavior is shown schematically in the inlet.

As mentioned before, microemulsions stabilized by a non-ionic amphiphile exhibit the phase sequence $\underline{2} \rightarrow 1 \rightarrow \overline{2}$. In this case, however, at temperatures below the one-phase region a lamellar L_α-phase coexists with a microemulsion phase (L_α present, see Figure 3.8). At lower temperatures a multiphase region containing a lamellar phase may exist (compare [194, 195]). To what extent the measured $\tilde{X}$-point is a "real" $\tilde{X}$-point is therefore unknown. At high temperatures a surfactant-rich water-in-CO_2 (w/c)-microemulsion phase coexists with a water excess phase ($\overline{2}$). As described in section 2.1.2, this phase sequence results from the curvature inversion of the amphiphilic film with increasing temperature. Increasing the pressure from $p = 150$ bar to $p = 200$ bar the $\tilde{X}$-point shifts to lower surfactant mass fractions. The pressure-dependence of the location of the $\tilde{X}$-point ($\tilde{\gamma}$ and $\tilde{T}$) of the balanced microemulsion is shown in Figure 3.4.

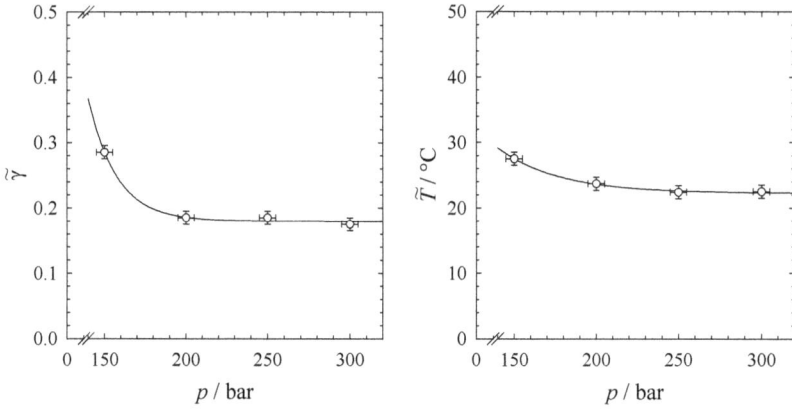

Figure 3.4: Surfactant mass fraction $\tilde{\gamma}$ (left) and temperature $\tilde{T}$ (right) at the optimum point $\tilde{X}$ of the balanced microemulsion system $H_2O/NaClO_4 - CO_2 - $ Capstone® FS-3100 at $\alpha = 0.40$ and $\varepsilon = 0.05$ as a function of the pressure p. As the pressure increases, both $\tilde{\gamma}$ and $\tilde{T}$ decay to a plateau value of $\tilde{\gamma} \approx 0.18$ and $\tilde{T} \approx 22$ °C. The experimental errors amount to $\Delta p = \pm 5$ bar, $\Delta\tilde{\gamma} = \pm 0.01$ and $\Delta\tilde{T} = \pm 1$ K.

As the pressure increases, both the surfactant mass fraction $\tilde{\gamma}$ and temperature $\tilde{T}$ at the optimum point $\tilde{X}$ decay to a plateau value. The efficiency of the surfactant Capstone® FS-3100 to solubilize $H_2O/NaClO_4$ and CO_2 strongly increases with increasing pressure, especially between $p = 150$ bar ($\tilde{\gamma} = 0.285$) and $p = 200$ bar ($\tilde{\gamma} = 0.185$). This can be explained with the increasing density of CO_2 at higher pressures (compare section 3.2.1, Figure 3.21) and thus a better solvatization of the polyfluorinated surfactant tails by CO_2. As a consequence, also the volume of the hydrophobic part of the surfactant increases and thus affects the curvature of the amphiphilic film. This results in a shift of the phase inversion temperature to lower temperatures. As the upper and lower phase boundaries are shifted simultaneously, the temperature extent of the one-phase region is unaffected by the pressure (see Figure 3.3). These findings are in agreement with the results from *Klostermann* for the Zonyl®-containing microemulsion system [30, 41].

iii) **CO_2-rich microemulsions**

Traversing the phase inversion, the phase behavior of the CO_2-rich microemulsion system was determined finally. Therefore, $T(w_A)$-sections through the phase prism were recorded starting from the binary $CO_2 - $ surfactant system at a constant surfactant mass fraction γ_b in the

CO_2/surfactant mixture. By the addition of water, defined by the overall water mass fraction w_A, the number and types of coexisting phases were determined as a function of temperature at different pressures. Figure 3.5 shows the recorded $T(w_A)$-sections of the CO_2-rich microemulsion system H_2O/$NaClO_4$ – CO_2 – Capstone® FS-3100 at $\gamma_b = 0.25$, $\varepsilon = 0.05$ and pressures of $p = 200$, 250 and 300 bar. To account for the high monomeric solubility of polyfluorinated surfactants in CO_2 the surfactant mass fraction was adjusted to $\gamma_b = 0.25$. *Klostermann* found a monomeric solubility of the Zonyl® surfactants in scCO_2 of up to $\gamma_{C,mon,b} \approx 0.17$ [30].

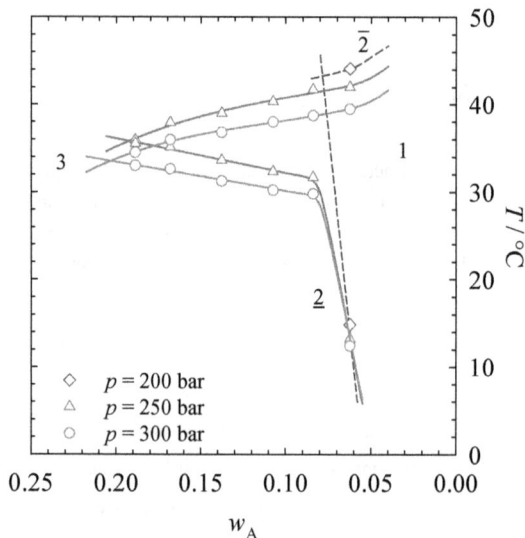

Figure 3.5: $T(w_A)$-phase diagrams of the CO_2-rich microemulsion system H_2O/$NaClO_4$ –CO_2 – Capstone® FS-3100 at $\varepsilon = 0.05$ and pressures of $p = 200$, 250 and 300 bar. Note, that a surfactant mass fraction of $\gamma_b = 0.25$ was used. Upon a pressure increase from $p = 200$ bar to $p = 250$ bar the efficiency of the surfactant to solubilize H_2O/$NaClO_4$ in CO_2 increases strongly. A further increase of the pressure only slightly increases the efficiency.

As can be seen in Figure 3.5, at low mass fractions of water w_A an extended one-phase region is found at intermediate temperatures. However, by increasing w_A the lower phase boundary $\underline{2} \to 1$ (*ncb*) ascends first strongly, then considerably slower, but always monotonically. The water emulsification failure boundary (*wefb*) $1 \to \overline{2}$ separates the one-phase region (1) from the $\overline{2}$-region, where a surfactant-rich water-in-CO_2 (w/c) microemulsion phase coexists with a water-excess phase $\overline{2}$. The *wefb* descends moderately and intersects the *ncb* at the upper critical end point temperature T_u and the maximum water mass fraction $w_{A,max}$. At larger values of w_A a three-phase region (3) is observed. Note, that in comparison to $p = 250$ and 300 bar significantly less water can be solubilized within the microemulsion at $p = 200$ bar. Inversely to the water-rich side, maximal swollen spherical water-in-CO_2 droplets are found along the $wefb$ [74]. Strongly structured microemulsions show a maximum in the ncb, which leads to the appearance of an additional closed-loop, i.e. a two-phase region in the $Gibbs$ triangle. The fact, that no maximum is observed in the CO_2-microemulsion might indicate that the microemulsion is only moderately structured or might be a consequence of the high salt concentration as discussed for the water-rich side.

The pressure-dependence of the maximum water mass fraction $w_{A,max}$ and the position of the upper temperature T_u are illustrated in Figure 3.6.

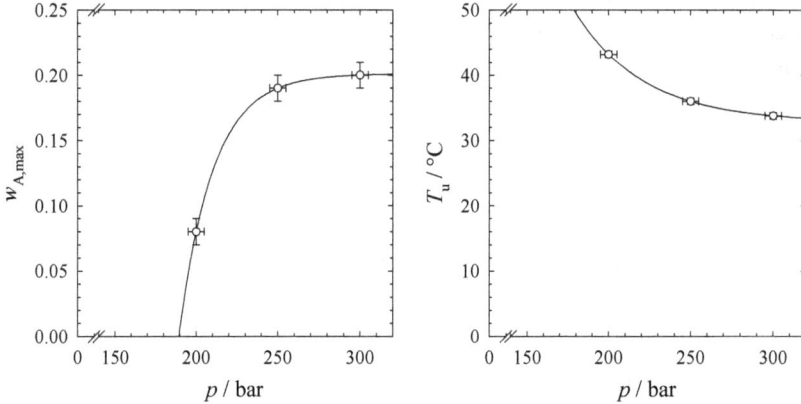

Figure 3.6: Maximum water mass fraction $w_{A,max}$ (left) and upper critical end point temperature T_u (right) of the CO_2-rich microemulsion system $H_2O/NaClO_4 - CO_2 -$ Capstone® FS-3100 at $\gamma_b = 0.25$ and $\varepsilon = 0.05$ as a function of the pressure p. With increasing pressure, the maximum water mass fraction $w_{A,max}$ increases, while the upper temperature T_u decreases. The experimental errors amount to $\Delta p = \pm 5$ bar, $\Delta w_{A,max} = \pm 0.01$ and $\Delta \tilde{T} = \pm 0.5$ K.

With increasing pressure the maximum water mass fraction $w_{A,max}$ increases, i.e. the $CO_2 -$ surfactant mixture becomes increasingly efficient in solubilizing water (see Figure 3.5 and Figure 3.6, left). At the same time, the upper critical end point temperature T_u monotonically shifts to lower temperatures. Thereby the intercept of ncb and $wefb$ shifts from $w_{A,max} = 0.08$ and $T_u = 43.2$ °C at $p = 200$ bar to $w_{A,max} = 0.20$ and $T_u = 33.8$ °C at $p = 300$ bar. Again both findings result from the corresponding CO_2-densities, which increase with increasing pressure (ρ_{CO_2}(200 bar, 43.2 °C) = 0.756 g·cm^{-3} to ρ_{CO_2}(300 bar, 33.8 °C) = 0.934 g·cm^{-3}) [29].

Thus, the polyfluorinated surfactant tails are better solubilized by CO_2-molecules and the overall volume of the hydrophobic surfactant tail increases. Consequently, the phase boundaries shift to lower temperatures to compensate this effect (see section 2.1.2).

3.1.2   Microstructure of CO_2-microemulsions elucidated by SANS

In order to probe the microstructure of the microemulsion system H_2O/NaClO$_4$ – CO_2 – Capstone® FS-3100 systematic small angle neutron scattering (SANS) studies were performed at the D11 instrument at the Institut Laue-Langevin in Grenoble, France. Furthermore, H_2O was replaced by D_2O in all samples in order to increase the scattering length density contrast and decrease the incoherent scattering intensity. Thereby, due to the fact that the exact molecular structure and composition of the surfactant Capstone® FS-3100 was unknown also its exact scattering length density was unknown and was therefore used as further fit parameter. Note, that in order to avoid confusion, the composition parameters used in the SANS part of this work are related to H_2O instead of D_2O, while marked with the index p.

i)   Water-rich microemulsions

The microstructure of the water-rich microemulsion of the type D_2O/NaClO$_4$ – CO_2 – Capstone® FS-3100 ($\gamma_{a,p}$ = 0.08, ε_p = 0.05) was studied at $w_{B,p}$ = 0.116 and pressures of p = 150, 200 and 300 bar. The temperature was adjusted to T = 5.2 °C, which is the lowest temperature adjustable in the stroboscopic high-pressure SANS cell (section 6.3.2). Thus, the SANS-measurements were performed in the one-phase region above the CO_2-*emulsification failure boundary* (*cefb*). While spherical CO_2-in-water droplets should exist at the *cefb*, increasing temperature at a given overall oil mass fraction w_B and pressure p is expected to result in a decrease of the curvature of the amphiphilic film and therewith to an elongation of the droplets. Thus, with increasing distance from the *cefb*, an increasing elongation of the CO_2-droplets is expected. Figure 3.7 shows the pressure-dependent scattering data obtained for the water-rich sample at T = 5.2 °C. For a better discrimination of the scattering curves, the data were multiplied by appropriate factors (10^x). As expected, the scattering data show the typical features of polydisperse cylindrical structures. At intermediate q-values the intensity decays with q^{-1}, followed by a q^{-4}-decay with a characteristic minimum followed by a maximum in the high q-part [54, 170], which indicates the existence of structures with relatively monodisperse cross-sectional radius. Additionally, at low q-values the intensity decreases stronger than q^{-1}, which might be attributed to critical fluctuations due to the proximity of the critical line cep_β. As can be seen in Figure 3.7, the position ($q \approx \pi/R_0$) and the form of the characteristic minimum/maximum at large q are independent of the applied pressure. Thus also the mean cross-sectional radius R_0 and polydispersity should be rather constant. However, with decreasing pressure the critical scattering becomes less pronounced indicating a larger distance from the critical line.

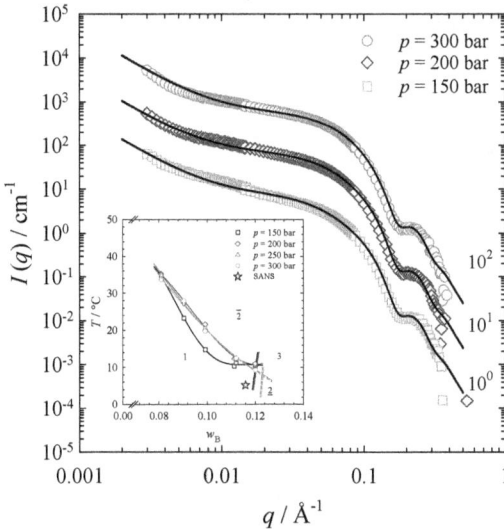

Figure 3.7: Scattering curves of a water-rich sample of the microemulsion system $D_2O/NaClO_4 - CO_2 - Capstone®$ FS-3100 with $\gamma_{a,p} = 0.08$, $\varepsilon_p = 0.05$ and $w_{B,p} = 0.116$ recorded at $T = 5.2 °C$ and pressures of $p = 150$, 200 and 300 bar. The experimental data are almost quantitatively described using the form factor of polydisperse cylinders $P_{cylinder}$ [51, 172, 173] in combination with the *Ornstein-Zernike* S_{OZ} [166, 179] and *Percus-Yevick* S_{PY} [176, 177] structure factors. Note that the different SANS-spectra have been discriminated by appropriate factors.

The scattering curves are almost quantitatively described using the polydisperse cylinder form factor $P_{cylinder}(q)$ (eq. 2.68) in combination with the *Ornstein-Zernike* $S_{OZ}(q)$ and *Percus-Yevick* (hard sphere approximation) $S_{PY}(q)$ structure factor (eq. 2.77). The structural parameters obtained from the analysis of the curve are listed in Table 3.1.

Table 3.1: Structural parameters (mean radius R_0, droplet polydispersity index σ/R_0, cylinder length L, Ornstein-Zernike correlation length ξ_{OZ} and Ornstein-Zernike parameter $S_{OZ}(0)$) obtained from the analysis ($n \cdot P_{cylinder}(q) \cdot S_{OZ}(q) \cdot S_{PY}(q)$ [51, 166, 172, 173, 179]) of the pressure dependent SANS-measurements of the water-rich microemulsion system $D_2O/NaClO_4 - CO_2 -$ Capstone® FS-3100 ($\gamma_{a,p} = 0.08$, $\varepsilon_p = 0.05$) at $w_{B,p} = 0.116$ and $T = 5.2$ °C. Note that, $\chi = 1$ Å, $d = 0$ Å, $\Delta\rho_{core} = 4.0 \cdot 10^{-6}$ Å^{-2}, $\Delta\rho_{film} = 4.0 \cdot 10^{-6}$ Å^{-2}, $v_c/a_c = 11$ Å, $\phi_{C,i} = 0.05$, $\phi_{disp} = 0.1$, $R_{HS} = R_0 + 20$ Å and $\sigma_{HS}/R_{HS} = 0.3$. With the sharpness parameter of the interface χ, the thickness of the surfactant layer d, the scattering length density contrasts $\Delta\rho_{core}$ and $\Delta\rho_{film}$, the approximate length of the surfactant molecule v_c/a_c, the interfacial volume fraction of surfactant $\phi_{C,i}$, the dispersed volume fraction ϕ_{disp}, the hard-sphere radius R_{HS} and the polydispersity index σ_{HS}/R_{HS}

p / bar	R_0 / Å	σ/R_0	L / Å	ξ_{OZ} / Å	$S_{OZ}(0)$
150	21	0.16	95	3000	600
200	21	0.16	80	4000	900
300	21	0.16	65	3500	900

As expected, a pressure independent mean cross-sectional radius of $R_0 = 21 \pm 3$ Å with a polydispersity index of $\sigma/R_0 = 0.16$ is obtained. Moreover, the length of the cylindrical structure decreases from $L = 95$ Å at $p = 150$ bar to $L = 65$ Å at $p = 300$ bar. From the decrease of the cylinder length upon pressure increase one may conclude, that the difference between $T = 5.2$ °C and the *cefb* at $w_{B,p} = 0.116$ decreases with increasing pressure. Note, that due to the high CO_2-density $\rho_{CO_2} \approx 1$ g·ml^{-1} at $T = 5.2$ °C the volume fractions and the scattering length density of the CO_2-droplet core were set to be constant.

The ratio v_c/a_c is approximately the length of the surfactant molecule l_C and was assumed to be $l_C = 11$ Å, which is slightly shorter than the length for the Zonyl® surfactants of $l_C = 14$ Å estimated by *Klostermann et al.* [41]. Note that the Zonyl® surfactants are in average $F(CF_2)_8C_2H_4E_j$, while Capstone® FS-3100 are presumable $F(CF_2)_6C_2H_4E_j$, i.e. the polyfluorinated surfactant tails are by C_2F_4 shorter. Assuming that the C-C bond length is 154 pm [128], the shortening of l_C by 3 Å seems to be reasonable.

The Zonyl®-containing system ($\varepsilon_p = 0.01$, $\delta_{FSN} = 0.75$) with $w_{B,p} = 0.10$ showed spherical CO_2-swollen micelles with a much greater mean radius of $R_0 = 66$ Å at $p = 150$ bar and cylinders with $R_0 = 48$–57 Å and $L = 230$–280 Å at $p = 200$–300 bar [46]. However, these systems were measured at $T = 30.0$ °C. The monomeric solubility of CO_2 in water increases from ~5.9 wt% at $p = 300$ bar and $T = 40$ °C to ~7.6 wt% at $p = 300$ bar and $T = 12$ °C [196]. Thus, at $T = 5.2$ °C more CO_2 is monomerically solubilized in water and consequently less CO_2 has to be dissolved in the cylinders, which is why the cross-sectional radius is smaller than for the Zonyl®-containing system.

ii) Balanced microemulsions

Having studied the microstructure of the water-rich microemulsion the microstructure of the balanced system $D_2O/NaClO_4 - CO_2 - $ Capstone® FS-3100 ($\alpha_p = 0.40$, $\varepsilon_p = 0.05$) was studied as a function of the pressure and surfactant mass fraction γ_p. For $p = 150$, 200 and 250 bar the measurements were performed in the middle of the one-phase region of the $T(\gamma)$-section. At $p = 300$ bar an additional measurements was performed below the one-phase region at $\gamma_p = 0.30$ and $T = 17.9$ °C, where a lamellar L_α-phase in coexistence with a microemulsion phase was detected (L_α present) in the phase behavior studies (see Figure 3.3). The respective SANS spectra are shown in Figure 3.8.

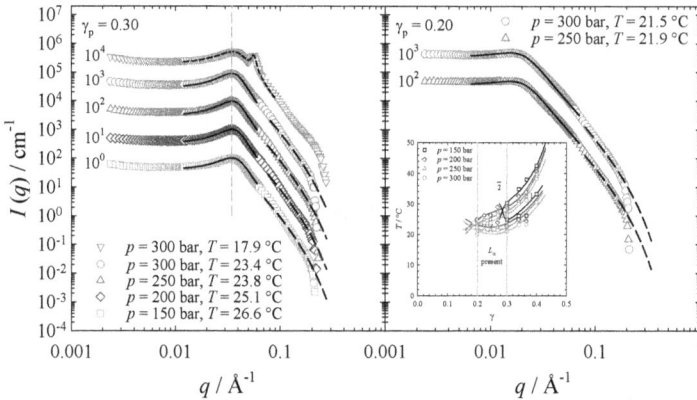

Figure 3.8: Bulk scattering curves of the balanced microemulsions $D_2O/NaClO_4 - CO_2 - $ Capstone® FS-3100 at $\alpha_p = 0.40$, $\varepsilon_p = 0.05$ and $\gamma_p = 0.30$ (left) and 0.20 (right). The SANS-curves were recorded at $p = 150$, 200, 250 and 300 bar, while the temperature was adjusted to the middle of the one-phase region. The data of the peak and high q-region were described using the *Teubner-Strey* [103] model (solid line) and the *Porod* [158, 159] decay for diffuse interfaces (dashed line), respectively. In addition the scattering curve recorded in the lower two-phase region (L_α present) at $\gamma_p = 0.30$, $p = 300$ bar and $T = 17.9$ °C is shown (left). In order to describe the two peaks a combination of two *Teubner-Strey* [103, 197] structure factors is used (short dashed line). The inlet shows the corresponding $T(\gamma)$-phase diagrams with the SANS-compositions (orange lines).

In each SANS-curve a scattering peak at intermediate q-values is observed, while the scattering intensity at low q-values is almost constant. At higher q-values the scattering intensity decays according to $\exp(-q^2 t^2) \cdot q^{-4}$, i.e. a *Porod* decay for diffuse interfaces [158, 159]. The shape of the scattering curves in combination with the phase behavior measurements supports the formation of a bicontinuous microstructure (see section 2.3.1) [86, 198]. The q-value of the scattering peak is directly correlated to the characteristic periodicity d_{TS} of the microstructure by $q_{max} \approx 2\pi/d_{TS}$. Decreasing the surfactant mass fraction γ shifts the scattering peak to lower q-values, indicating a structure with larger periodicity d_{TS}. Furthermore, the peak broadens with decreasing γ, which indicates a weakening of the structural order due to undulations of the amphiphilic film. The scattering curve recorded within the lower two-phase region at $\gamma_p = 0.30$, $p = 300$ bar and $T = 17.9$ °C show an additional peak at $q \approx 0.056$ Å^{-1}. The existence of two peaks, a broad and a narrow one, are a strong indication for the coexistence of a lamellar and bicontinuous phase [197, 199, 200]. Note, that an isotropic signal was recorded, while a comparable system showed an anisotropic signal (compare Figure 3.22). Moreover, in all scattering curves a shoulder is observed at $2q_{max}$, resulting from multiple scattering due to the rather large sample thickness of 2 mm given by the SHP-SANS cell.

The *Teubner-Strey* model [103, 197]

$$I(q) = \frac{1}{a_2 + c_1 q^2 + c_2 q^4}$$

3.1

was applied to describe the scattering peaks of the samples from the one-phase region (solid lines). As described before (see section 2.3.1), the periodicity d_{TS}, correlation length ξ_{TS} and amphiphilicity factor f_a can be determined from the parameters a_2, c_1 and c_2. In order to describe both scattering peaks observed in the scattering curve of the ME+L_α-sample

$$I_{ME+L_\alpha}(q) = \frac{1}{a_2 + c_1 q^2 + c_2 q^4} + \frac{1}{a_{2,L_\alpha} + c_{1,L_\alpha} q^2 + c_{2,L_\alpha} q^4}$$

3.2

was used [197]. The high q-part of the scattering data were evaluated using the *Porod* decay for diffuse interfaces (section 2.3.1) [158, 159]

$$\lim_{q \to \infty} [I(q)] = 2\pi \frac{\Delta\rho_s^2}{q^4} \frac{S}{V} \exp\left(-q^2 t^2\right) .$$

3.3

Thus the specific internal interface S/V and the volume fraction of surfactant at the interface can be determined. The parameters obtained from the analysis of the intermediate (*Teubner-Strey* model) and high q-region (*Porod*) of the scattering curves are listed in Table 3.2 and Table 3.3, respectively.

Table 3.2: Characteristic periodicity d_{TS}, correlation length ξ_{TS} and amphiphilicity factor f_a resulting from the *Teubner-Strey* parameters [103] from the analysis of the scattering peaks of the balanced microemulsion $D_2O/NaClO_4 - CO_2 - $ Capstone® FS-3100 ($\alpha_p = 0.40$, $\varepsilon_p = 0.05$) at various surfactant mass fractions γ_p, pressures p and temperatures T. Note that, the scattering data of the sample at $\gamma_p = 0.30$, $p = 300$ bar and $T = 17.9$ °C (coexistence region ME+L_α) was described by the combination of two *Teubner-Strey* [103, 197] structure factors (see eq. 3.2).

γ_p	p / bar	T / °C	d_{TS} / Å	ξ_{TS} / Å	f_a	d_{TS,L_α} / Å	ξ_{TS,L_α} / Å	f_{a,L_α}
0.30	150	26.6	171	79.8	-0.792	--	--	--
0.30	200	25.1	172	88.4	-0.825	--	--	--
0.30	250	23.8	173	91.8	-0.835	--	--	--
0.30	300	23.4	174	92.1	-0.835	--	--	--
0.30	300	17.9	168	77.2	-0.786	112	304	-0.993
0.20	250	21.9	342	94.3	-0.500	--	--	--
0.20	300	21.5	344	98.3	-0.528	--	--	--

The structural parameters listed in Table 3.2 confirm the trends from the qualitative discussion of the scattering curves. By decreasing the surfactant mass fraction γ the periodicity strongly increases from $d_{TS} = 174$ Å at $\gamma_p = 0.30$ to $d_{TS} = 344$ Å at $\gamma_p = 0.20$ (both at $p = 300$ bar and similar temperature). Simultaneously, the amphiphilicity factor becomes less negative from $f_a = -0.835$ to -0.528 indicating the effect of stronger undulations. As expected, the specific internal interface S/V decreases with decreasing surfactant mass fraction from $S/V = 0.0197$ Å^{-1} at $\gamma_p = 0.30$ to $S/V = 0.0105$ Å^{-1} at $\gamma_p = 0.20$ (see Table 3.3). Considering the parameters obtained from the analysis of the scattering curve recorded in the ME+L_α region at $\gamma_p = 0.30$, $p = 300$ bar and $T = 17.9$ °C, smaller values of $d_{TS} = 168$ Å, $\xi_{TS} = 77.2$ Å and a less negative value of f_a are found. Analyzing the sharp peak, an amphiphilicity factor of $f_{a,L_\alpha} = -0.993$ is found. Being close to -1, this result proves the existence of a lamellar phase (L_α).

The comparison of the periodicities of the two phases yields a ratio of $d_{TS}/d_{TS,L_\alpha} = 1.497$, which is very close to the predicted value. Explaining this correlation more precisely, the characteristic length scale ξ of a bicontinuous structure is predicted by several theoretical models

$$\xi = a \frac{a_c}{v_c} \frac{\phi(1 - \phi)}{\phi_{c,i}} .$$

3.4

Here the prefactor a depends on the model used. While *Debye et al.* propose $a = 4$ [161], *Talmon* and *Prager* obtained $a = 5.82$ using a *Voronoi* tessellation [160] and *De Gennes* and

Taupin predict $a = 6$ [80]. Experimentally *Sottmann* and *Strey* found $a = 7.0$ [56]. With $d_{TS}/\xi = 2$ [86], the thickness of the amphiphilic film $v_C/a_C = l_C$ and the volume fraction $\phi = 0.388$ (adjusted in the respective sample) one can calculate a periodicity (eq. 3.4 with $a = 6$) of $d_{TS} = 2.849 \cdot l_C/\phi_{C,i}$. For the lamellar microstructure the periodicity can be calculated by $d_{TS,L_\alpha} = 2 \cdot l_C/\phi_{C,i}$. Thus, neglecting the waste of interface due to undulation in both cases, a ratio $d_{TS}/d_{TS,L_\alpha} = 1.425$ is obtained from these geometrical considerations, which is in good agreement with the experimentally observed value of $d_{TS}/d_{TS,L_\alpha} = 1.497$.

Decreasing the pressure at $\gamma_p = 0.30$ from $p = 300$ bar to $p = 150$ bar the periodicity d_{TS} slightly decreases from $d_{TS} = 174$ Å to $d_{TS} = 171$ Å, while the correlation length decreases considerably from $\xi_{TS} = 92.1$ Å to $\xi_{TS} = 79.8$ Å. Accordingly, the amphiphilicity factor becomes less negative, i.e. increases from $f_a = -0.835$ to $f_a = -0.792$. Thus, a decreasing pressure leads to a weaker structuring of the sample, which is in agreement with the results from *Klostermann* [41]. It turned out, that $\ln(\tilde{\gamma})$ is proportional to the CO_2-density, which increases with increasing pressure and decreasing temperature (see Figure 3.21) [41].

The volume fraction of surfactant at the interface $\phi_{C,i}$ can be calculated from the specific internal interface S/V by $\phi_{C,i} = l_C \cdot S/V$ [56, 158]. As pointed out before (Figure 3.7), the length of the surfactant molecule for Capstone® FS-3100 was assumed to be $l_C = 11$ Å.

As can be seen in Table 3.3, S/V slightly decreases with increasing pressure. Consequently, also the volume fraction of surfactant at the interface is reduced, which presumably results from a slight increase of the monomeric solubility of surfactant in CO_2 with increasing pressure.

Using the interfacial volume fraction of surfactant $\phi_{C,i}$, the volume fraction of surfactant ϕ_C and the volume fraction of CO_2 ϕ_{CO_2}, the volume fraction of surfactant monomeric solubilized in CO_2 can be calculated by

$$\phi_{C,mon} = \phi_C - \phi_{C,i} \qquad\qquad 3.5$$

and

$$\phi_{C,mon,b} = \frac{\phi_{C,mon}}{\phi_{C,mon} + \phi_{CO_2}} . \qquad\qquad 3.6$$

Thus, depending on pressure p and surfactant mass fraction γ relatively large monomeric solubilities of Capstone® FS-3100 in CO_2 of $\phi_{C,mon,b} = 0.03-0.13$ are found (see Table 3.3). While the increase of $\phi_{C,mon,b}$ with decreasing surfactant mass fraction is unexpected, this might be caused by a depletion zone of surfactant close to the amphiphilic film at higher surfactant mass fractions. Moreover, the length of the surfactant is unknown and was therefore assumed to be $l_C = 11$ Å. The uncertainty of l_C will contribute to the $\phi_{C,mon,b}$-value as well as to the difference of the values at $\gamma = 0.20$ and 0.30. While similar values were described for the monomeric solubility $\gamma_{C,mon,b}$ of the Zonyl® surfactants in CO_2 (water – CO_2 – $F(CF_2)_iC_2H_4E_j$) [30], for H_2O – n-alkane – C_iE_j significantly lower monomeric solubilities of non-ionic C_iE_j surfactants in n-alkanes $\gamma_{C,mon,b} \approx 0.01$-$0.06$ were determined [70].

Interestingly, following the model of *Safran* and *Pieruschka* [156], the *Teubner-Strey* parameters d_{TS} and ξ_{TS} allows to calculate the renormalized bending rigidity κ_{SANS} via

$$\frac{\kappa_{SANS}}{k_B T} = \frac{10\sqrt{3}\pi}{64} \frac{\xi_{TS}}{d_{TS}} . \qquad 3.7$$

Please note that due to new models (theoretical considerations) by *Gompper et al.* [201], which are confirmed by *Sottmann and Holderer et al.* [202], the form of the peak (given by d_{TS} and ξ_{TS}) is determined not only by κ_{SANS} but also by the renormalized saddle splay modulus $\kappa_{0,SANS}$. As the ratio ξ_{TS}/d_{TS} strongly depends on the surfactant mass fraction, this trend is also observed for κ_{SANS}. By increasing the surfactant mass fraction at $p = 300$ bar from $\gamma_p = 0.20$ to $\gamma_p = 0.30$, the renormalized bending rigidity increases from $\kappa_{SANS} = 0.25$ $k_B T$ to $\kappa_{SANS} = 0.45$ $k_B T$. Furthermore, it slightly increases with increasing pressure from $\kappa_{SANS} = 0.40$ $k_B T$ at $p = 150$ bar to $\kappa_{SANS} = 0.45$ $k_B T$ at $p = 300$ bar (both at $\gamma = 0.30$). Although Capstone® FS-3100 is a shorter chain fluorinated surfactant, the values of the renormalized bending rigidity are in agreement with values obtained for the Zonyl®-stabilized CO_2-microemulsions [41, 114]. However, the renormalized bending rigidity contains all membrane fluctuations with wavelengths up to a length scale of the present structural dimensions, which leads to a considerable softening of the membrane. The renormalization corrected bare bending rigidity $\kappa_{0,SANS}$ is related to κ_{SANS} according to [92, 93, 155]

$$\frac{\kappa_{0,SANS}}{k_B T} = \frac{\kappa_{SANS}}{k_B T} + \frac{3}{4\pi} \ln\left(\frac{d_{TS}}{2l_C}\right) . \qquad 3.8$$

Determining the renormalization corrected bending rigidity by this relation, the stiffness of the amphiphilic film yields values of $\kappa_{0,SANS} \approx 0.90$–$0.95$ $k_B T$ (see Table 3.3), which correspond almost quantitatively to values found for other CO_2-microemulsions stabilized by Zonyl® surfactants [41, 114].

Table 3.3: Scattering length density contrast $\Delta\rho_s$, specific internal interface S/V, diffusivity of the amphiphilic film t, volume fraction of surfactant ϕ_C, interfacial volume fraction of surfactant $\phi_{C,i}$ and volume fraction of surfactant monomerically solubilized in CO_2 $\phi_{C,mon,b}$ resulting from the *Porod*-analysis [158, 159] of the high q-range parts of the scattering spectra of the balanced microemulsion system $D_2O/NaClO_4$ - CO_2 – Capstone® FS-3100 ($\alpha_p = 0.40$, $\varepsilon_p = 0.05$) at various surfactant mass fractions γ_p, pressures p and temperatures T. The renormalized bending rigidity κ_{SANS} and renormalization corrected bending rigidity $\kappa_{0,SANS}$ were calculated using equation 3.7 and 3.8, respectively.

γ_p	p / bar	T / °C	$\dfrac{\Delta\rho_s /}{10^{-6}\,Å^{-2}}$	S/V / $Å^{-1}$	t / $Å$	ϕ_C	$\phi_{C,i}$	$\phi_{C,mon,b}$	$\dfrac{\kappa_{SANS} /}{k_B T}$	$\dfrac{\kappa_{0,SANS} /}{k_B T}$
0.30	150	26.6	4.1	0.0204	6.2	0.234	0.224	0.029	0.397	0.886
0.30	200	25.1	4.1	0.0201	6.0	0.238	0.221	0.052	0.436	0.928
0.30	250	23.8	4.1	0.0199	5.8	0.241	0.219	0.069	0.452	0.944
0.30	300	23.4	4.1	0.0197	5.6	0.243	0.217	0.081	0.451	0.944
0.20	250	21.9	4.0	0.0105	4.5	0.157	0.116	0.123	0.243	0.894
0.20	300	21.5	4.0	0.0105	4.3	0.158	0.116	0.130	0.248	0.896

The pressure dependence of the renormalized and renormalization corrected bending rigidity obtained for the amphiphilic film of balanced microemulsions of the type $D_2O/NaClO_4 - CO_2 -$ Capstone® FS-3100 ($\alpha_p = 0.40$, $\varepsilon_p = 0.05$) at $\gamma_p = 0.20$ and 0.30 are shown in Figure 3.9.

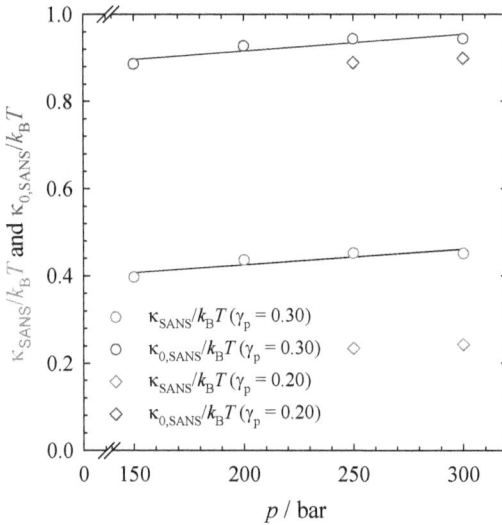

Figure 3.9: Renormalized bending rigidity κ_{SANS} (red) and renormalization corrected bending rigidity $\kappa_{0,SANS}$ (blue) obtained for the amphiphilic film of the balanced (bicontinuous microstructure) microemulsion $D_2O/NaClO_4 - CO_2 -$ Capstone® FS-3100 ($\alpha_p = 0.40$, $\varepsilon_p = 0.05$) at $\gamma_p = 0.20$ (diamonds) and 0.30 (circles) as a function of the pressure p at the respective temperatures (see Table 3.3).

Interestingly, but expected, accounting for the thermal undulation at $\gamma_p = 0.20$ and 0.30 the values of the renormalization corrected bending rigidity are almost independent of the surfactant mass fraction γ. However, the trend that with increasing pressure κ_{SANS} slightly increases is also observed for $\kappa_{0,SANS}$. Note, that the influence of a low-molecular, hydrophobic co-oil, e.g. cyclohexane, on the bending rigidities κ_{SANS} and $\kappa_{0,SANS}$ of balanced CO_2-microemulsions is studied in section 3.2.

iii) CO_2-rich microemulsions

As next step SANS-measurements were performed to fully characterize the microstructure of water-in-CO_2-microemulsions. Having determined the monomeric solubilities of the surfactant in CO_2 of $\phi_{C,mon,b} = 0.03{-}0.13$ for the bicontinuous system (see Table 3.3), a rather high surfactant mass fraction of $\gamma_b = 0.25$ was adjusted. The corresponding phase behavior is shown in Figure 3.5. For the SANS-measurements the temperatures were adjusted slightly below the upper *we/b*, where maximal swollen spherical water-in-CO_2 droplets are expected. In Figure 3.10 the scattering curves of the studied system are shown as a function of pressure ($w_{A,p}$ = 0.075, left) and water mass fraction $w_{A,p}$ (p = 300 bar, right).

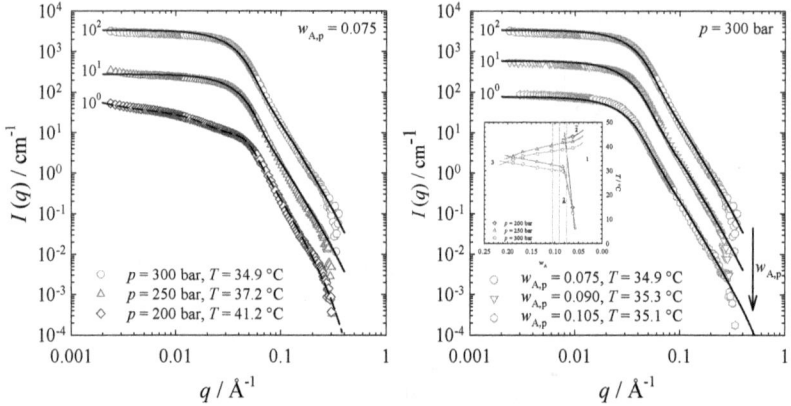

Figure 3.10: Scattering curves of the CO_2-rich microemulsion $D_2O/NaClO_4 - CO_2 -$ Capstone® FS-3100 at γ_b = 0.25, ε_p = 0.05 recorded at different $w_{A,p}$-values and pressures, respectively. Temperatures just below the *wefb* were adjusted. *Left*: pressure-dependence at $w_{A,p}$ = 0.075, *right*: $w_{A,p}$-dependence at p = 300 bar. The experimental data are described using a combination of the droplet form factor for polydisperse spheres with only core contributions $P_{droplet}$ [51, 170, 203] and the *Percus-Yevick* [176, 177] structure factor S_{PY} (solid line). The scattering data at $w_{A,p}$ = 0.075 and p = 200 bar were fitted using the polydisperse cylinder form factor $P_{cylinder}$ [51, 172, 173] in combination with the *Ornstein-Zernike* [166, 179] and *Percus-Yevick* structure factor dashed line). Note, that the different SANS-spectra have been multiplied by appropriate factors for discrimination.

The scattering spectra recorded at p = 250 and 300 bar and $w_{A,p}$ = 0.075, 0.090 and 0.105, show the typical scattering pattern of polydisperse spherical particles. At low q-values the scattering intensity is constant and decreases steadily with q^{-4} at higher q-values. Note, that almost no dent can be observed at intermediate, respectively high q-values, indicating a relatively large polydispersity. For even larger q-values a $\exp(-q^2t^2) \cdot q^{-4}$-decay of the scattering intensity indicates, that the scattering length density profile is not a sharp step profile but rather a diffuse profile. The scattering data were described using a combination of the form factor of polydisperse droplets $P_{droplet}$ using the scattering length density profile suggested by *Foster et al.* [51] and *Percus-Yevick* structure factor S_{PY}. Please note, that the scattering length densities of CO_2 and Capstone® FS-3100 are close to $2 \cdot 10^{-6} Å^{-2}$. Thus the scattering length density differences between CO_2 and the amphiphilic film were estimated to be $\Delta\rho_{film} \approx 0 Å^{-2}$, i.e. only contributions from the core were considered. The scattering curve at $p = 200$ bar, $T = 41.2$ °C and $w_{A,p} = 0.075$ shows a q^{-1}-decay at intermediate q-values, while at medium to large q-values the already described $\exp(-q^2 t^2) \cdot q^{-4}$-decay occurs. Again the scattering intensity decays monotonically, indicating a relatively large polydispersity. Additionally, an increase of the scattering intensity at very low q-values can be assigned to critical fluctuations. As already described for the microstructure of the water-rich system, a q^{-1}-decay is a typical feature for elongated droplets and cylinders. Hence, the scattering data at $p = 200$ bar were described using the form factor of polydisperse cylinder $P_{cylinder}$ in combination with the *Percus-Yevick* S_{PY} as well as *Ornstein-Zernike* structure factor S_{OZ}. The structural parameters obtained from the analysis of the scattering curves are compiled in Table 3.4.

Table 3.4: The structural parameters mean radius R_0, polydispersity σ/R_0 and cylinder length L of the form factor of polydisperse droplets $P_{droplet}$ [51, 170, 203] and cylinders $P_{cylinder}$ [51, 172, 173] obtained from the analysis of the scattering data and scattering length density difference $\Delta\rho_{core}$ of the system $D_2O/NaClO_4 - CO_2 -$ Capstone® FS-3100 ($\gamma_b = 0.25$, $\varepsilon_p = 0.05$) at various $w_{A,p}$, p and T values. The diffusivity of the amphiphilic film was adjusted $t = 2$ Å, while for $p = 200$ bar $t = 4$ Å was found. Moreover the *Percus-Yevick* [176, 177] (hard-sphere diameter d_{HS}, volume fraction of dispersed phase ϕ_{disp}) and *Ornstein-Zernike* [166, 179] (correlation length of critical fluctuations ξ_{OZ}, *Ornstein-Zernike* parameter S_{OZ}) structure factors were applied.

$w_{A,p}$	p / bar	T / °C	R_0 / Å	σ / R_0	L / Å	$\Delta\rho_{core}$ / 10^{-6}Å^{-2}	d_{HS}/ Å	ϕ_{disp}	ξ_{OZ} / Å	$S_{OZ}(0)$
0.075	200	41.2	41	0.40	300	-4.16	90	0.215	1000	3
0.075	250	37.2	41	0.41	--	-4.07	86	0.073	--	--
0.075	300	34.9	43	0.44	--	-3.97	90	0.077	--	--
0.090	250	36.3	43	0.44	--	-4.07	90	0.088	--	--
0.090	300	35.3	47	0.47	--	-4.02	98	0.090	--	--
0.105	250	36.2	49	0.41	--	-4.07	102	0.104	--	--
0.105	300	35.1	51	0.47	--	-4.02	106	0.106	--	--

A mean radius of $R_0 = 41$ to 51 Å, with a polydispersity index of $\sigma/R_0 \approx 0.40$–0.47 was obtained from the fits. The mean radius R_0 increases linearly with increasing water content w_A, which is in agreement with geometrical considerations (see section 2.1.2) and observations by *Eastoe et al.* [48, 49]. In addition the radius as well as polydispersity slightly increases with increasing pressure and decreasing temperature. Thereby the increase in radius can be related to the increasing monomeric solubility of the surfactant in CO_2 with increasing pressure and decreasing temperature. *Eastoe et al.* determined core radii of $R_c = 12$ to 36 Å with $\sigma/R_c \approx 0.20$

for water-in-CO_2 droplet microemulsions using a partially fluorinated anionic surfactant at $T = 15$ and 25 °C, respectively [48, 49]. For another fluorinated anionic surfactant similar radii of $R_c = 15$ to 21 Å were described by *Lee et al.* [204]. Moreover, *Lee et al.* have shown that the water droplet radius depends on the density of CO_2, while the radius is slightly increasing with increasing pressure. The higher polydispersity indices σ/R_0 found in this work, as compared to *Eastoe et al.* [48, 49], might result due to the fact that the water-in-CO_2 microemulsions were formulated at higher temperatures of $T = 34.1$ to 41.2 °C, where the density of CO_2 is lower and thus the interaction between the fluorinated surfactant chain and CO_2 is weaker.

The scattering curve at $w_{A,p} = 0.075$, $T = 41.2$ °C and $p = 200$ bar is almost quantitatively described by the polydisperse cylinder fit-curve in combination with the two structure factors. Thus, polydisperse cylinders with a mean radius of $R_0 = 41$ Å, a polydispersity index of $\sigma/R_0 = 0.40$ and a length of $L = 300$ Å can be found. The elongation of the structure as well as the existence of critical fluctuations is caused by the proximity of the critical endpoint cep_α.

3.1.3 Corresponding state description for the length scales found in CO_2-microemulsions

As pointed out before, so far a scaling of the most prominent features was only found for microemulsion systems of the type water – n-alkane – C_iE_j and water – n-alkane – C_iG_j (alkylpolyglycoside) [53, 54, 56]. *Klostermann et al.* could show that the trajectory of the middle phase of CO_2-microemulsion systems resembles that of classical n-alkane microemulsion systems. Thus, he was able to scale the trajectory of both systems on one single curve. Furthermore, he studied the characteristic length scale of the water-rich and balanced water – $CO_2 - F(CF_2)_iC_2H_4E_j$ system [41, 45]. Although he wasn't able to characterize the microstructure of the w/c-microemulsion, also the temperature dependence of the characteristic length scale of CO_2- and classical microemulsions exhibit a strong similarity.

Having studied in this thesis the structural evolution from c/w- over bicontinuous to w/c-microemulsions allows comparing the temperature dependence of the characteristic length scale of CO_2 and classical microemulsions. Some definitions beforehand for balanced bicontinuous microemulsions the characteristic length scale was defined by *Strey* as $\xi = d_{TS}/2$ [54], while the characteristic length scale of spherical droplets is given by the mean radius R_0. In section 3.1.2 the microstructure of the water-rich, balanced as well as of the CO_2-rich microemulsion of the type $D_2O/NaClO_4 - CO_2 - $ Capstone® FS-3100 was determined by means of SANS. In order to proof, whether the corresponding state description [53, 54, 56] for water –

n-alkane – C_iE_j microemulsions is applicable to CO_2-microemulsions, the reduced characteristic length scale $\xi \cdot \phi_{C,i,m}$ and the reduced temperature $2(T\text{-}T_m)/T_u\text{-}T_l$ are required. T_u and T_l can each be determined from the intersection of the *efb* and *ncb* (see section 2.1.2). However, for the system investigated in this thesis, this is expected to be a crude approximation. Technical grade non-ionic fluorinated surfactants, e.g. the used Capstone® FS-3100, exhibit a broad distribution in the ethoxylation degree. Especially the surfactant molecules with lower ethoxylation degrees are extracted into the CO_2. Furthermore, the absolute amount of extracted surfactant molecules increases with increasing CO_2 volume fraction. Consequently, the amphiphilic film becomes increasingly more hydrophilic. While for the water-rich $T(w_B)$-section this effect is expected to be negligible as the mass fraction of CO_2 is low, it strongly affects the properties of balanced and in particular CO_2-rich microemulsions. As a consequence, the phase boundaries are shifted to higher temperatures if diluted with oil (and water) as shown for the microemulsion system H_2O – n-octane – technical C_iE_j (Figure 2.6). Thus, the extracted $\Delta T = T_u\text{-}T_l$ from the phase diagrams will therefore produce too large values for microemulsions stabilized by technical surfactants exhibiting a broad distribution e.g. in the ethoxylation degree.

For microemulsion systems of the type water – n-alkane – C_iE_j it is found that the interfacial volume fraction of surfactant at the optimum point $\phi_{C,i,m}$ and the width of the three-phase region are correlated not only for ternary systems approaching the tricritical endpoint, but also for well-structured microemulsion systems [55, 56]. For both types of systems it turned out, that $\phi_{C,i,m}$ increases almost linearly with increasing ΔT (see Figure 3.11, left). Assuming, that this trend also holds true for CO_2-microemulsion systems, ΔT was calculated from the interfacial volume fraction of surfactant $\phi_{C,i,m}$.

Please note, that $\phi_{C,i,m}$ was calculated by

$$\phi_{C,i,m} \approx \phi_C - \phi_{C,mon,b} \cdot \phi_{CO_2} \qquad\qquad 3.9$$

with the CO_2 volume fraction ϕ_{CO_2}, the volume fraction ϕ_C of surfactant and the volume fraction $\phi_{C,i,m}$ of surfactant monomerically solubilized in CO_2, which are listed in Table 3.3. The in this way determined values are summarized in Table 3.5.

Table 3.5: Calculated volume fraction ϕ_{CO_2} of CO_2, interfacial volume fraction $\phi_{C,i,m}$ of surfactant at the optimum point $\tilde{X}$ and temperature width ΔT calculated thereof.

p / bar	γ	ϕ_{CO_2}	$\phi_{C,i,m}$	ΔT / K
150	0.30	0.321	0.194	12.9
200	0.30	0.310	0.193	12.9
250	0.20	0.335	0.103	6.9
300	0.20	0.330	0.102	6.8

In the next step both the characteristic length scale and temperature are reduced to $\xi \cdot \phi_{C,i,m}$ and $2(T-T_m)/\Delta T$ and afterwards plotted. The reduced characteristic length scales $\xi \cdot \phi_{C,i,m}$ of the microemulsion system $H_2O/NaClO_4 - CO_2 -$ Capstone® FS-3100 ($\varepsilon = 0.05$) in comparison with $H_2O - n$-octane $- C_{10}E_4$ [55, 56] and $H_2O/NaCl - CO_2 -$ Zonyl® FSN 100/FSO 100 ($\delta_{FSN} = 0.75$, $\varepsilon = 0.01$) [45] are shown in Figure 3.11 as function of the reduced temperature. For the fits see equations 2.18 and 2.19.

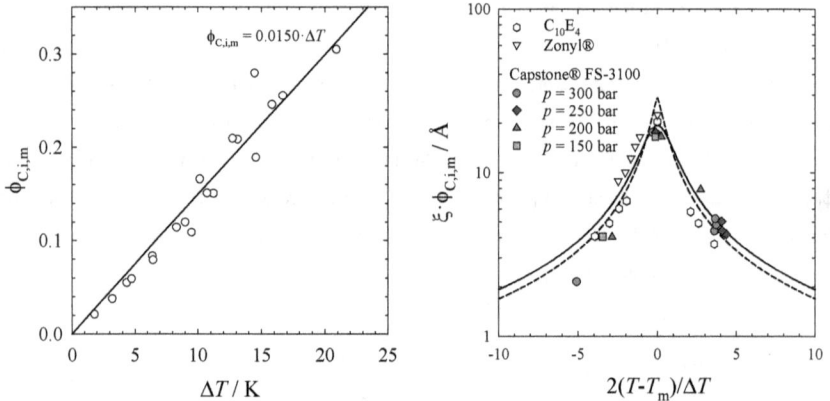

Figure 3.11: *Left*: Interfacial volume fraction of surfactant at the optimum point $\phi_{C,i,m}$ as a function of the temperature width of the three phase body ΔT for microemulsion systems of the type $H_2O - n$-octane $- C_iE_j$ [55, 56]. *Right*: Reduced characteristic length scale $\xi \cdot \phi_{C,i,m}$ as function of the reduced temperature $2(T-T_m)/\Delta T$ for $H_2O - n$-octane $- C_{10}E_4$ (hollow hexagons) [55, 56], $H_2O/NaCl - CO_2 -$ Zonyl FSN 100/FSO 100 (hollow triangles) [45] and $H_2O/NaClO_4 -$ $CO_2 -$ Capstone® FS-3100 at $p = 150$, 200, 250 and 300 bar (filled symbols). Note, that due to a distorted phase behavior of the system $H_2O/NaClO_4 - CO_2 -$ Capstone® FS-3100 the ΔT-values of this system were extrapolated from the $\phi_{C,i,m}(\Delta T)$ (left). The fits were calculated for the system $H_2O - n$-octane $- C_{12}E_5$ from equations 2.18 (solid line, [54]) and 2.19 (dashed line, [56]), while the parameters where taken from [55, 86, 106].

It becomes obvious that, as found for the oily counterparts [56], the length scale of the structure runs through a maximum at the phase inversion temperature. Moreover, the data points collapse almost into a single curve with only one data point obtained for the cylindrical CO_2-in-water microemulsion deviating. Thus, these results clearly indicate that the corresponding state description, found by *Sottmann* and *Strey* [53, 54, 56] for classical state of the art microemulsions containing alkanes, is also applicable to CO_2-microemulsion systems.

3.1.4 Dynamics of bicontinuous CO_2-microemulsions studied by NSE

So far the bending rigidity of the balanced microemulsion system was indirectly determined from SANS-measurements (Figure 3.9). In the following, the bending rigidity is directly determined via neutron spin echo (NSE) measurements by characterizing the dynamics of the amphiphilic film fluctuations. A short overview of the fundamentals on NSE is given in section 2.3.2. The measurements were performed using the stroboscopic high pressure SANS cell, while the experimental setup used is presented in section 6.3.4.

NSE-measurements were performed on the system $D_2O/NaClO_4 - CO_2 - $ Capstone® FS-3100 using the same compositions ($\alpha_p = 0.40$, $\varepsilon_p = 0.05$, $\gamma_p = 0.20$ and 0.30), pressures and temperatures as adjusted for the SANS-measurements. Note, that the composition parameters are related to the protonated system (index p). Figure 3.12 exemplary shows the normalized intermediate scattering function $S(q,\tau)/S(q,0)$ recorded at $\gamma_p = 0.30$ and $p = 150$ bar as function of the *Fourier* time τ. Note, that the normalized intermediate scattering function $S(q,\tau)/S(q,0)$ was measured at 15 different q-values between $q = 0.0273$ Å^{-1} and $q = 0.1182$ Å^{-1} (3 different detector positions recording the $S(q,\tau)/S(q,0)$ of 5 q-values at the same time). For better clarity only $S(q,\tau)/S(q,0)$-curves of 6 q-values are shown. Moreover, as the pressure transducer was connected to the IN15 instrument, for each *Fourier* time τ the current pressure was recorded. The pressure as function of the *Fourier* time τ and for the three detector positions measured is plotted in Figure 3.12 on top.

As can be seen, during a NSE-measurement (which lasts about 4.5 h) the pressure slightly decreases. In total, for this measurement, the pressure drops by about $\Delta p = 5.3$ bar, which corresponds to ~ 1.2 bar/h.

In a first evaluation step, the experimental data were fitted using the *Zilman-Granek* model, i.e. $S(q,\tau)/S(q,0) = \exp\left(-\left(\Gamma_{ZG}(q) \cdot \tau\right)^{\beta_{ZG}}\right)$ (solid lines in Figure 3.12), which is valid in the high q-regime (see section 2.3.2). The *Zilman-Granek* relaxation rate Γ_{ZG} is defined as [114]

$$\Gamma_{ZG}(q) = 0.025 \gamma_k \left(\frac{k_B T}{\kappa}\right)^{1/2} \frac{k_B T}{\eta_0} q^3 \qquad\qquad 3.10$$

with

$$\gamma_k = 1 - 3\left(\frac{k_B T}{4\pi\kappa}\right) \ln(q\xi_{TS}) . \qquad\qquad 3.11$$

Thereby, Γ_{ZG} and the stretching exponent β_{ZG} were used as the fitting parameters. As can be seen, the monoexponential function describes the NSE-data qualitatively, but not quantitatively (see e.g. intermediate *Fourier* times for $q = 0.0543$–0.0919 Å^{-1}).

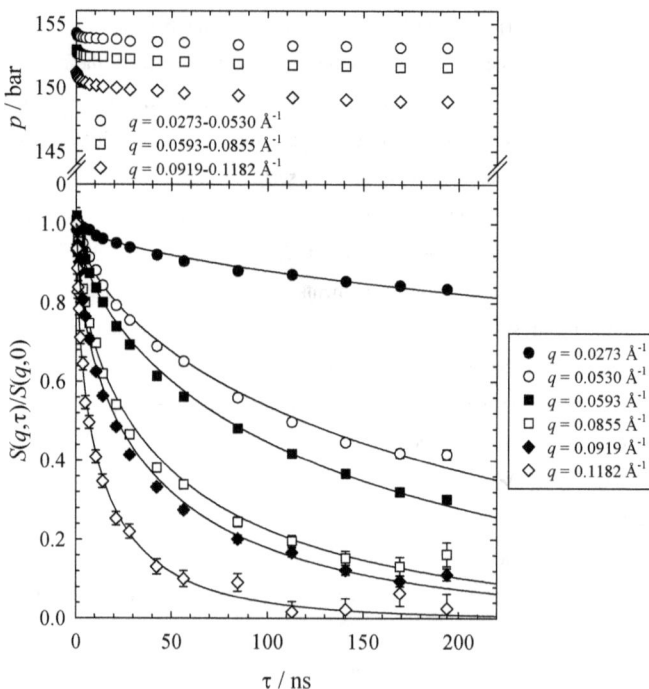

Figure 3.12: Exemplary normalized intermediate scattering functions $S(q,\tau)/S(q,0)$ of the bicontinuous microemulsion D_2O/NaClO$_4$ – CO_2 – Capstone® FS-3100 at $\alpha_p = 0.40$, $\varepsilon_p = 0.05$, $\gamma_p = 0.30$, $p = 150$ bar and $T = 26.6$ °C as function of the *Fourier* time τ. The solid lines show the fits according to the representation of the *Zilman-Granek* model $S(q,\tau)/S(q,0) = \exp\left(-\left(\Gamma_{ZG}(q)\cdot\tau\right)^{\beta_{ZG}}\right)$. Simultaneously to the data acquisition, the current pressure was measured, indicating a decrease of the pressure by ~ 1.2 bar/h (top).

The fitting parameters Γ_{ZG} and β_{ZG} for the system at $\gamma_p = 0.20$ and 0.30 and various pressures are plotted in Figure 3.13 as function of q^3. While the stretching exponent β_{ZG} only slightly oscillate around the predicted value of $\beta_{ZG} = 2/3$ [185], the scaling law $\Gamma_{ZG} \sim q^3$ (eq. 3.10) suggested by *Zilman* and *Granek* does not describe the data at low q-values. However, the scaling law seems to be fullfilled at high q-values. Comparing the relaxation rates of bicontinuous microemulsions with different surfactant mass fractions γ, a considerably smaller rate is found for the $\gamma_p = 0.30$ microemulsion. Furthermore, with increasing pressure a slight decrease of the relaxation rate is observed. The deviations from the predicted q^3-dependence in the low-q regime are most probably a consequence of the proximity to the peak position (see Figure 3.8). Thus, the relaxation rate contains a large contribution stemming from the translational diffusion of the water and CO_2 domains. Furthermore, according to the de Gennes narrowing, the data evaluation is strongly affected by the structure factor [205, 206]. There the relaxation rate is $\Gamma(q) \sim \Gamma_D q^2 \cdot S(q)^{-1}$, with Γ_D being the translational diffusion relaxation rate of the domains and $S(q)$ being the *Teubner-Strey* structure factor [185].

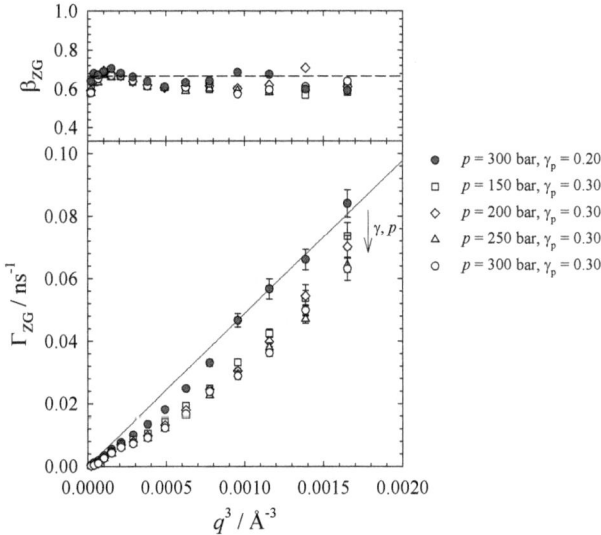

Figure 3.13: *Zilman-Granek* relaxation rate Γ_{ZG} as function of q^3 obtained from evaluation of the NSE-data of the bicontinuous microemulsion $D_2O/NaClO_4 - CO_2 -$ Capstone® FS-3100 ($\alpha_p = 0.40$, $\varepsilon_p = 0.05$) at $\gamma_p = 0.20$ and $p = 300$ bar and $\gamma_p = 0.30$ at $p = 150, 200, 250$ and 300 bar according to eq. 2.83. In addition, the stretching exponent β_{ZG} is shown on top. The dashed line in $\beta_{ZG}(q)$ represents the theoretically predicted value $\beta_{ZG} = 2/3$.

Using eq. 3.9 and 3.10, the bending rigidity κ_{NSE} is determined and shown in Figure 3.14 as function of q. Additionally, the bending rigidity values for the system $D_2O/NaCl - CO_2 -$ Zonyl FSN 100/FSO 100 ($\alpha_p = 0.40$, $\gamma_p = 0.262$, $\delta_{FSN} = 0.40$, $\varepsilon_p = 0.01$, $p = 300$ bar, $T = 31.1$ °C), which were determined by *Klostermann*, are shown in comparison [45]. As can be seen, *Klostermann et al.* studied the dynamics of the system on a wider q-range, i.e. at a higher q-range and thus minimized contributions from translational diffusion.

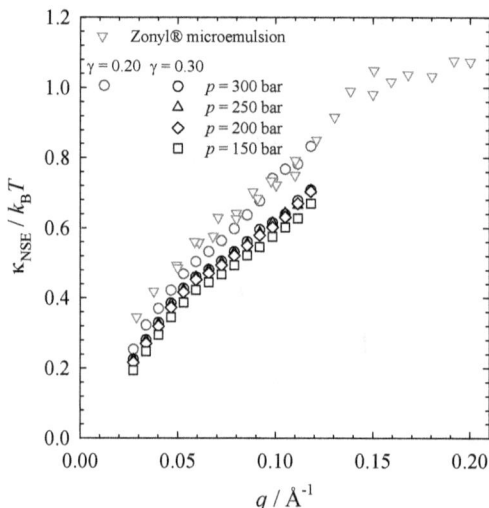

Figure 3.14: Bending rigidity κ_{NSE}/k_BT of the bicontinuous microemulsion system $D_2O/NaClO_4 - CO_2 -$ Capstone® FS-3100 ($\alpha_p = 0.40$, $\varepsilon_p = 0.05$) at $\gamma_p = 0.20$ ($p = 300$ bar) and $\gamma_p = 0.30$ and $p = 150$, 200, 250 and 300 bar obtained from the *Zilman-Granek* relaxation rate Γ_{ZG} via eq. 3.10 as a function of the scattering vector q. For comparison, also the bending rigidity of the system $D_2O/NaCl - CO_2 -$ Zonyl FSN 100/FSO 100 ($\alpha_p = 0.40$, $\gamma_p = 0.262$, $\delta_{FSN} = 0.40$, $\varepsilon_p = 0.01$, $p = 300$ bar, $T = 31.1$ °C) determined by *Klostermann* [45] is shown.

For the Zonyl®-stabilized system, at low q-values, a linear increase of the bending rigidity with q becomes obvious, while a plateau value of $\kappa_{NSE} \approx 1.05\ k_BT$ is formed at $q > 0.15$ Å^{-1} [30]. The Capstone®-containing systems also show the increase of the bending rigidity with increasing q. However, in the investigated q-range ($q = 0.0273-0.1182$ Å^{-1}) κ_{NSE} does not reach a plateau.

The deviations of the relaxation rate from the q^3-dependence and also the increase of the bending rigidity at low q-values can be related to translational diffusion (hydrodynamic collective motions) of water or CO_2 domains, which are so far not included in the analysis [207]. Note, that most of the measurements were unfortunately performed within the intermediate q-range or even at $q < q_{max}$, where the diffusion modes are dominating [185, 208]. In order to include contributions of both local undulations (*Zilman-Granek* model) and collective motions the normalized intermediate scattering function can be described by [209, 210]

$$\frac{S(q,\tau)}{S(q,0)} = a\exp\left(-Dq^2\tau\right)\left[\left(1-A_{ZG}\right)+A_{ZG}\cdot\exp\left(-\left(\Gamma_{ZG}(q)\cdot\tau\right)^{\beta_{ZG}}\right)\right].$$ 3.12

With the prefactor a, the translational diffusion coefficient D ($\Gamma_D = D\cdot q^2$), A_{ZG} being the amplitude of the *Zilman-Granek* contributions. However, as described by *Arriaga et al.* [209], no stable fits are obtained from this equation, while the linear combination of two independent processes

$$\frac{S(q,\tau)}{S(q,0)} = A_D\exp\left(-\Gamma_D\tau\right)+\left(1-A_D\right)\exp\left(-\left(\Gamma_{ZG}(q)\cdot\tau\right)^{\beta_{ZG}}\right)$$ 3.13

yields stable fits, with A_D being the amplitude of the diffusion process. Note that eq. 3.13 is an approximation which is only valid for systems, where the time scales of diffusion and fluctuations are different. Having analyzed the normalized intermediate scattering function according to eq. 3.12, the obtained relaxation rates Γ_D and Γ_{ZG} for the system $D_2O/NaClO_4 - CO_2 -$ Capstone® FS-3100 ($\alpha_p = 0.40$, $\varepsilon_p = 0.05$) are shown in Figure 3.15 as function of q^2 and q^3, respectively.

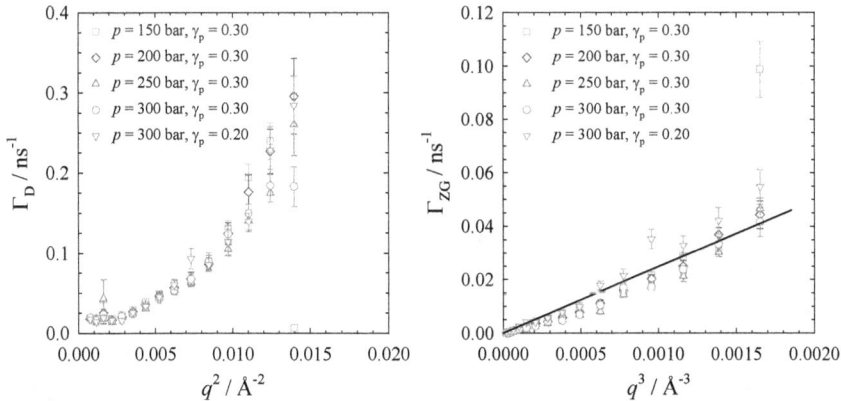

Figure 3.15: Translational diffusion relaxation rate Γ_D as function of q^2 (left) and fluctuation domianted (*Zilman-Granek*) relaxation rate Γ_{ZG} as function of q^3 (right) of the bicontinuous microemulsion $D_2O/NaClO_4 - CO_2 -$ Capstone® FS-3100 ($\alpha_p = 0.40$, $\varepsilon_p = 0.05$) at $\gamma_p = 0.30$ and $p = 150$, 200, 250 and 300 bar and $p = 300$ bar at $\gamma_p = 0.20$. The relaxation rates were obtained from the analysis of the NSE-data according to eq. 3.13 [209].

As can be seen, Γ_{ZG} increase almost linearly with q^3, indicating that this relaxation rate is fluctuation dominated. However, $\Gamma_D(q^2)$ has a parabolic shape, indicating that this relaxation rate contains not only contributions from translational diffusion. An explanation for this result are the similar time scales, of the both processes represented in Γ_D especially at low q, which makes it difficult to separate both contributions from each other. As a consequence also the amplitude of the diffusion mode A_D, which is expected to steadily decrease with increasing q, is found to increase from $A_D \approx 0.1$ at low q to $A_D \approx 0.3$–0.4 at intermediate and high q-values. Considering the trend of Γ_{ZG} with pressure and surfactant mass fraction, almost no trend of the relaxation rates with pressure is observed, while Γ_{ZG} increases slightly when γ_p is decreased from $\gamma_p = 0.30$ to 0.20.

Calculating the mean diffusion coefficient for the CO_2-microemulsion studied at $p = 300$ bar and $\gamma_p = 0.30$, a value of $<D> \approx 1.2 \cdot 10^{-10}$ m^2 s^{-1} is found. Using an average viscosity of $\eta_0 = [\eta_0(CO_2) + \eta_0(H_2O/NaClO_4)] / 2 = 6.2 \cdot 10^{-4}$ Pa·s can be determined via the *Stokes-Einstein* equation

$$D = \frac{k_B T}{6\pi\eta_0 R_h}$$

3.14

a "hydrodynamic radius R_h" of $R_h \approx 35$ Å for the domains of the bicontinuous microemulsion system. Hence, the value is only about a factor of 1.5 smaller than the radius of water/CO_2-domains. Keeping in mind, that the *Stokes-Einstein* equation is only valid for non-interacting diluted, spherical structures and not for balanced bicontinuous systems, this qualitative agreement is surprising.

As the *Zilman-Granek* relaxation rates show a q^3-dependence, they were analyzed using eq. 3.10 yielding new values of the bending rigidity κ_{NSE}, which are at least partly corrected for translational diffusion. The respective values are plotted in Figure 3.16 as function of q.

As can be seen, the bending rigidity $\kappa_{NSE}/k_B T$ still increases with increasing scattering vector q. However, the data can be described via an exponential increase yielding the maximum value $\kappa_{NSE,max}$ (long dashed lines) at high q-values. Interestingly, the bending rigidities obtained from NSE are in almost qualitative agreement with the renormalization corrected bending rigidity $\kappa_{0,SANS}$ (short dashed line), being just slightly smaller. As mentioned before, the renormalization corrected bending rigidity $\kappa_{0,SANS}$ contains some uncertainties in the length of the surfactant molecule l_C and thus of the renormalization factor $3/4\pi \cdot \ln(d_{TS}/(2 \cdot l_C))$ (see 2.3.1), while in the analysis of Γ_{ZG} the use of an average viscosity is an approximation.

Considering the variation of κ_{NSE} with varying pressures and surfactant mass fractions, only a small but systematic increase of $\kappa_{NSE,max}$ with increasing pressure from 0.85 k_BT at 150 bar to 0.87 k_BT at 300 bar is found for $\gamma_p = 0.30$. Furthermore, $\kappa_{NSE,max}$ increases with decreasing surfactant mass fraction γ_p from $\kappa_{NSE,max}/k_BT \approx 0.86$ at $\gamma_p = 0.30$ to $\kappa_{NSE,max}/k_BT = 0.96$ at $\gamma_p = 0.20$, both at 300 bar. This is somewhat unexpected (compare Figure 3.9), as the value κ_{NSE} should not depend on the surfactant mass fraction. However, with respect to all approximations made, the obtained κ_{NSE} values compare reasonable well the bending rigidity determined for amphiphilic films of classical n-alkane microemulsions [185, 211] and CO_2-microemulsions [41].

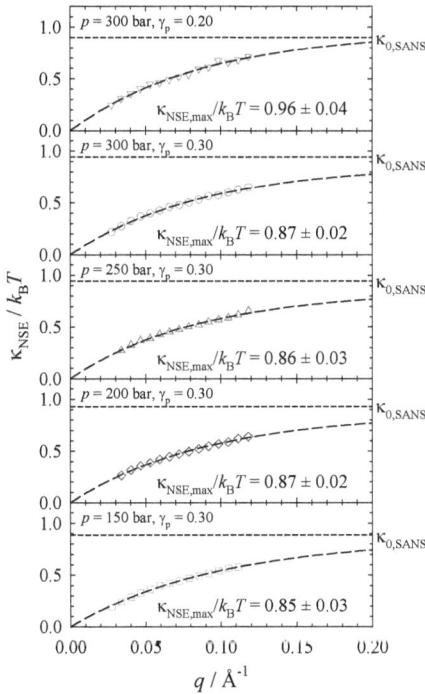

Figure 3.16: Bending rigidity κ_{NSE}/k_BT of the amphiphilic film in the bicontinuous microemulsion system $D_2O/NaClO_4 - CO_2 -$ Capstone® FS-3100 ($\alpha_p = 0.40$, $\varepsilon_p = 0.05$) at $\gamma_p = 0.30$ and $p = 150$, 200, 250 and 300 bar and $p = 300$ bar at $\gamma_p = 0.20$. The bending rigidity was determined from evaluation of the NSE-data with $S(q,\tau)/S(q,0) = A_D \cdot \exp(-\Gamma_D \cdot \tau)$ $\cdot (1 \cdot A_D) \cdot \exp(-(\Gamma_{ZG} \cdot \tau)^{2/3})$ and analyzing the *Zilman-Granek* relaxation rate Γ_{ZG} according to eq. 3.10. The bending rigidity increases exponentially to a maximum value $\kappa_{NSE,max}/k_BT$ (long dashed lines). For comparison the respective renormalized corrected bending rigidities obtained from SANS $\kappa_{0,SANS}$ are shown (short dashed line).

NSE-measurements on the Zonyl®-stabilized CO_2-microemulsion analyzed with a numerical integration of the *Zilman-Granek* model (see eq. 3.15) revealed a stronger increase of the bending rigidity κ_{NSE} with increasing pressure [41]. However, this is a main difference between the Zonyl®- and Capstone®-stabilized CO_2-microemulsion: while the former one was studied at supercritical conditions, i.e. at temperatures above the critical temperature of CO_2 ($T_c = 30.98$ °C), the one-phase region of the Capstone®-stabilized CO_2-microemulsion is located below $T_c(CO_2)$, at $T \approx 24$ °C. Thus, CO_2 is in liquid state exhibiting a higher density and viscosity as supercritical CO_2. Furthermore, the results go hand in hand with the phase behavior studies, where $\tilde{X}$ was found to depend only slightly on the pressure (except 150 bar).

A more quantitative and reliable determination of the bending rigidity using the *Zilman-Granek* model requires numerical integration of the intermediate scattering function over the different undulation modes. Hence, the NSE-data of the bicontinuous microemulsion system $D_2O/NaClO_4 - CO_2 -$ Capstone® FS-3100 at $\alpha_p = 0.40$, $\gamma_p = 0.30$, $\varepsilon_p = 0.05$, $p = 150$ bar and $T = 26.6$ °C were analyzed by

$$S(q,\tau) = \frac{2\pi\xi^2}{a^4} \int_0^1 d\mu \int_0^{r_{max}} dr\, r\, J_0\left(qr\sqrt{1-\mu^2}\right)$$
$$\cdot \exp\left(-\frac{k_B T}{2\pi\kappa} \cdot q^2\mu^2 \cdot \int_{k_{min}}^{k_{max}} dk\, \frac{1-J_0(kr)\cdot\exp(-\omega(k)\cdot\tau)}{k^3}\right) \qquad 3.15$$

For a more detailed description including the integration limits see section 2.3.2. Eq. 3.12 was used to account for translational diffusion. Thereby, the diffusion coefficient D was considerd as a fit parameter, but was kept constant for all q-values. Again the average viscosity of water and CO_2 was used. The obtained diffusion and *Zilman-Granek* relaxation rates Γ_D and Γ_{ZG} are plotted in Figure 3.17 as function of q^2 and q^3, respectively.

Due to the constant diffusion coefficient D, the diffusion relaxation rate Γ_D scales linearly with q^2, while the *Zilman-Granek* relaxation rate Γ_{ZG} scales linearly as function of q^3 with only slight deviations. However, the relaxation rates of both processes determined by this more quantitative analysis also differ only by a factor of ~ 2, i.e. they are on the same time scale.

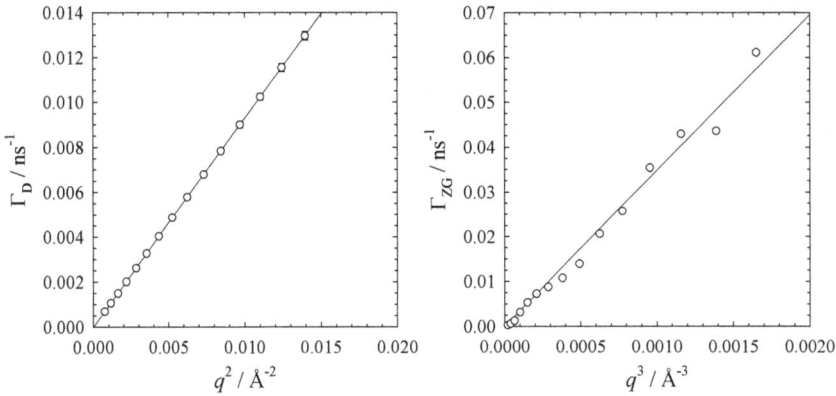

Figure 3.17: Translational diffusion relaxation rate Γ_D as function of q^2 (left) and Zilman-Granek relaxation rate Γ_{ZG} as function of q^3 (right) of the bicontinuous microemulsion $D_2O/NaClO_4 - CO_2 - $ Capstone® FS-3100 at $\alpha_p = 0.40$, $\gamma_p = 0.30$, $\varepsilon_p = 0.05$, $p = 150$ bar and $T = 26.6$ °C. The relaxation rates were obtained via the numerical integration of the intermediate scattering function (see eq. 3.15 and section 2.3.2) taking into account contributions from translational diffusion and assuming a constant diffusion coefficient, that yields in $D = 9.285 \cdot 10^{-12}$ m^2 s^{-1}.

Note that the analysis of the data reveals a diffusion coefficient of $D = 9.285 \cdot 10^{-12}$ m^2 s^{-1} and a hydrodynamic radius of $R_h \approx 420$ Å. The diffusion coefficient is by about one order of magnitude smaller than the value found from the previous analysis, where the hydrodynamic radius was found to be of the same order of magnitude. Due to the smaller diffusion coefficient, the hydrodynamic radius is one order of magnitude larger. Considering eq. 3.15 it becomes clear, that the lower relaxation rate of the diffusion process increases the relaxation rate of the fluctuation process and as a consequence leads to smaller κ_{NSE}/k_BT-values.

The amplitudes related to the diffusion mode A_D and the Zilman-Granek relaxation (fluctuation) mode A_{ZG} are shown together with the bending rigidity κ_{NSE}/k_BT in Figure 3.18.

In contrast to the evaluation using the simple Zilman-Granek approximation, the full numerical analysis (eq. 3.15) yields the expected result. While the amplitude of the diffusion mode is decreasing with increasing q from $A_D \approx 0.94$ to $A_D \approx 0.30$, the amplitude of the fluctuation related mode increases from $A_{ZG} \approx 0.06$ to $A_{ZG} \approx 0.70$ (Figure 3.18, left). However, the bending rigidity κ_{NSE}/k_BT (Figure 3.18, right) decreases simultaneously from $\kappa_{NSE}/k_BT - 0.61$ to $\kappa_{NSE}/k_BT - 0.10$, which was set as the lower limit of the model.

Thus, the bending rigidity is significantly lower than the values obtained from the analysis of the SANS-spectra, which are in agreement with the values obtained from analysis of the NSE-data using the simple *Zilman-Granek* approximation. Furthermore, the latter values of the bending rigidity are of the same order of magnitude than the values found for Zonyl®-stabilized CO_2-microemulsions.

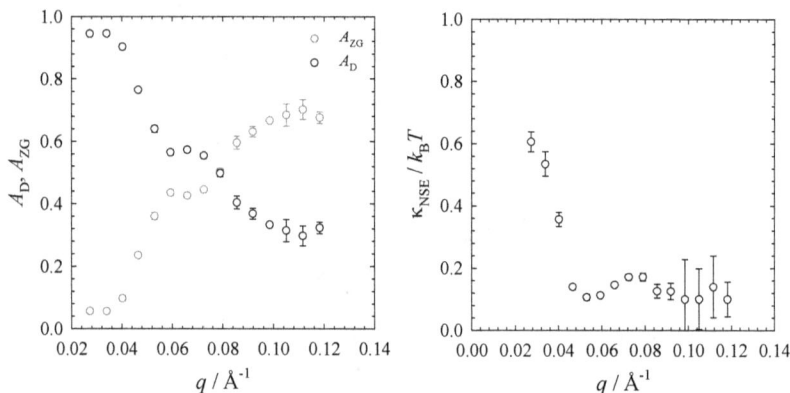

Figure 3.18: Amplitudes (left) of the diffusion mode A_D (black symbols) and *Zilman-Granek* relaxation mode A_{ZG} (red symbols) as well as bending rigidity $\kappa_{NSE}/k_B T$ (right) of the bicontinuous microemulsion $D_2O/NaClO_4 - CO_2 -$ Capstone® FS-3100 at $\alpha_p = 0.40$, $\gamma_p = 0.30$, $\varepsilon_p = 0.05$, $p = 150$ bar and $T = 26.6$ °C as function of q. Note that the values were obtained from numerical integration (see eq. 3.15 and section 2.3.2), while the diffusion coefficient was kept constant and the lower limit of the bending rigidity was set to $\kappa_{NSE}/k_B T = 0.1$.

In summary, the numerical evaluation suffers from the fact that both processes, the translational diffusion and the undulations of the amphiphilic film, are on the same time scale. Thus, a quite small diffusion coefficient $D = 9.285 \cdot 10^{-12}$ m^2 s^{-1} is obtained from the fits and no reliable values of the bending rigidity were extracted. A possible but sophisticated approach to overcome these challenges is to explicitly specify the diffusion coefficient via the structure factor. Therefore, for now the evaluation of the NSE-data with the simple *Zilman-Granek* approximation (eq. 3.13) seems to be more reliable.

3.2 Co-oil cyclohexane: efficiency boosting and Anti-Aging-Agent

The results presented in the previous section 3.1 clearly show the applicability of the corresponding state description for length scales of microemulsion systems of the type water $-$ $CO_2 - F(CF_2)_iC_2H_4E_j$. However, even though the Capstone® surfactants are less environmental harmful than previously used Zonyl® surfactants, a further reduction of the fluorine content is desirable. Furthermore, *Strey et al.* [39] suggested that the spinodal separation of CO_2 and a low-molecular, hydrophobic co-oil slow down the expansion-induced coarsening of the structure (compare 2.1.4). As an enrichment of the co-oil at the interface is proposed and due to the larger density of the co-oil and expected lower interfacial tension between H_2O and co-oil, an increase of the stability of the oil-domains is expected. For the microemulsion system of the type $H_2O/NaCl - CO_2 - Zonyl®$ FSO 100/FSN 100 ($\alpha = 0.40$, $\delta_{FSN} = 0.75$, $\varepsilon = 0.01$) *Pütz et al.* observed an efficiency boosting upon partial replacement of CO_2 by the co-oil cyclohexane [47]. In detailed studies she found, that the efficiency follows a pressure-specific parabolic trajectory as a function of the mass fraction of cyclohexane in the CO_2/cyclohexane mixture.[57] Thereby a reduction of the amount of surfactant needed to formulate a one-phase microemulsion by a factor of 2 to 5 was found. Sophisticated SANS-measurements on water-rich CO_2-microemulsions applying contrast variation showed, that due to repulsive interactions between the cyclohexane and the fluorinated hydrophobic surfactant part cyclohexane accumulated in the center of the CO_2-swollen micelles, avoiding the proximity of the fluorinated surfactant [47]. Furthermore, it was proposed that the depletion of cyclohexane close to the amphiphilic film might increase the bending rigidity of the amphiphilic film and therefore being the reason for the found efficiency boosting [58, 59].

In order to determine, whether the partial replacement of CO_2 by cyclohexane affects the bending rigidity of the amphiphilic film, neutron spin echo (NSE) measurements were performed in this thesis by means of the system $D_2O/NaClO_4 - CO_2$/cyclohexane $-$ Capstone® FS-3100. Thus, the phase behavior (section 3.2.1) and microstructure (SANS, section 3.2.2) were studied before the dynamics were examined by means of NSE (section 3.2.3). The obtained results are presented in the following sections.

3.2.1 Phase behavior

The phase behavior of the balanced microemulsion system $H_2O/NaClO_4 - CO_2$/cyclohexane $-$ Capstone® FS-3100 at $\alpha = 0.40$, $\beta = 0.20$, $\varepsilon = 0.05$ and pressures of $p = 150$, 200, 250 and

300 bar is shown in Figure 3.19 as function of the surfactant mass fraction γ and temperature T. Thus, 20 wt%, i.e. $\beta = 0.20$, of the CO_2 were replaced by the low molecular co-oil cyclohexane. In addition, the phase behavior of the cyclohexane-free system ($\beta = 0.00$) is shown for each investigated pressure for better comparison.

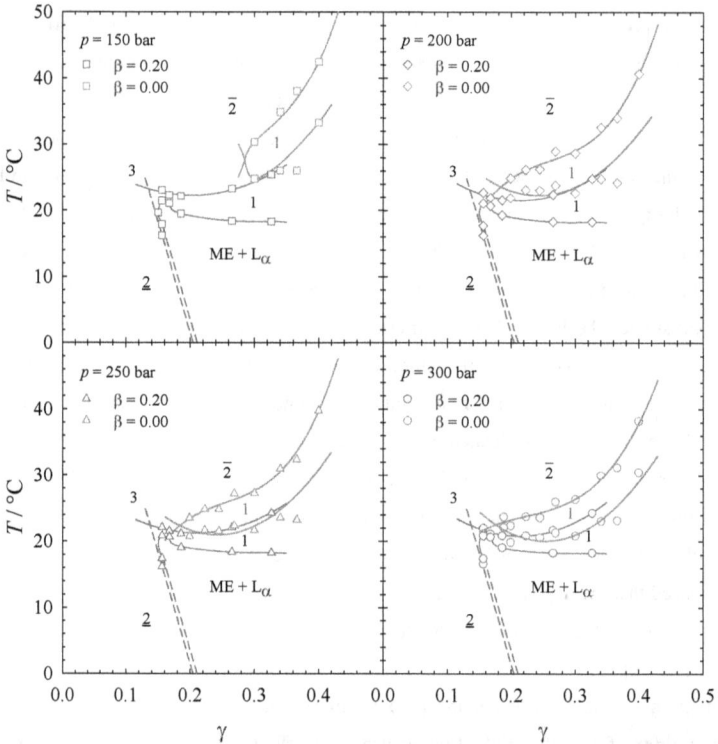

Figure 3.19: $T(\gamma)$-section through the phase prism with $\beta = 0.00$ (grey) as well as for $\beta = 0.20$ (red) of the system $H_2O/NaClO_4 - CO_2$/cyclohexane – Capstone® FS-3100 at $\alpha = 0.40$, $\varepsilon = 0.05$ and $p = 150$ to 300 bar and. With increasing pressure from 150 bar to 300 bar the upper phase boundaries slightly shift to lower temperatures. Below the one-phase region an almost pressure-independent coexistence region of a microemulsion phase with a lamellar phase (denoted as $ME+L_\alpha$) was found in the cyclohexane-containing system. Furthermore, in the cyclohexane-containing system a narrow isotropic one-phase channel was found at temperatures below the extended $ME+L_\alpha$ coexistence region.

As can be seen, the partial replacement of CO_2 by cyclohexane leads to a strong increase of efficiency at $p = 150$ bar, while for higher pressures ($p = 200$, 250 and 300 bar) the efficiency boosting is considerably smaller. For the Zonyl®-stabilized system a similar behavior, however more systematic behavior was found. Furthermore, as found for the cyclohexane-free system (β = 0.00, Figure 3.3 and Figure 3.8) and will be proven by SANS (Figure 3.22), an extended 2-phase region where a lamellar L_α-phase coexists with a bicontinuous phase (denoted ME+L_α) can be observed at low temperatures. However, in the β = 0.20 system at even lower temperatures an isotropic one-phase region is found between the ME+L_α-coexistence region and the $\underline{2}$-phase region, where a surfactant-rich oil-in-water (o/w) microemulsion phase coexists with an oil excess phase. Additionally, at lower surfactant mass fractions, i.e. close to the optimum point $\tilde{X}$, the ME+L_α-region closes and the complete phase sequence $\underline{2} \rightarrow 1 \rightarrow \overline{2}$ could be found, while in the cyclohexane-free system the phase inversion was not observed. This supports the assumption, that the $\tilde{X}$-point measured in the system without co-oil is not the "real" $\tilde{X}$-point (see Figure 3.3). At higher temperatures a surfactant-rich water-in-CO$_2$ (w/c) microemulsion phase and a water excess phase coexists, denoted as $\overline{2}$. Note, that the phase behavior of the cyclohexane-containing system is almost pressure-independent.

This pressure-independency of the $\tilde{X}$-point ($\tilde{\gamma}$ and $\tilde{T}$) of the system with β = 0.20 is illustrated in Figure 3.20. Independent of the pressure, a surfactant mass fraction of $\tilde{\gamma}$ $\approx$ 0.137 is required to form a one-phase CO$_2$-microemulsion. Compared to the system without cyclohexane (β = 0.00) the surfactant mass fraction is thus reduced by 24% to 48% (at p = 150 bar).

Considering the dependence of the phase inversion temperature $\tilde{T}$ on the pressure, it can be seen that $\tilde{T}$ is almost independent of the pressure p for both the cyclohexane-free and cyclohexane-containing system. Thus, $\tilde{T}$ converges towards $\tilde{T}$ $\approx$ 22.3 °C.

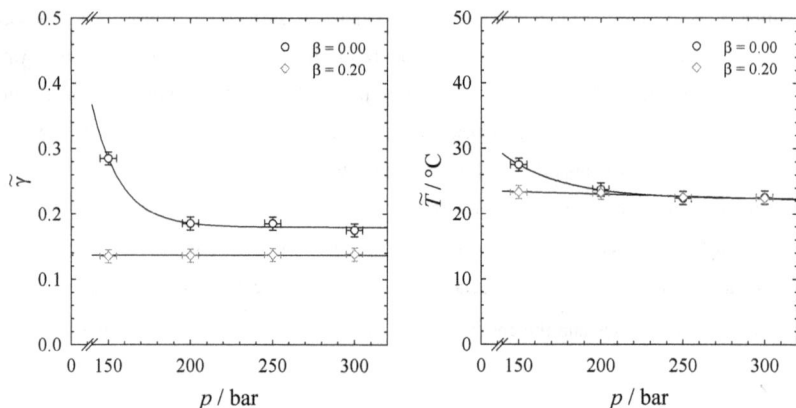

Figure 3.20: Surfactant mass fraction $\tilde{\gamma}$ (left) and temperature $\tilde{T}$ (right) at the optimum point $\tilde{X}$ of the balanced microemulsion system $H_2O/NaClO_4 - CO_2/cyclohexane -$ Capstone® FS-3100 at $\alpha = 0.40$, $\varepsilon = 0.05$ and $\beta = 0.20$ (red diamonds) as a function of pressure. The respective values for $\beta = 0.00$ (black circles) are shown for comparison. As the pressure increases, for $\beta = 0.00$ both values, $\tilde{\gamma}$ and $\tilde{T}$, decay to a plateau value, while constant values are observed for both values for the cyclohexane-rich system with $\beta = 0.20$. Interestingly, the addition of cyclohexane at $p = 150$ bar reduces $\tilde{\gamma}$ by about 48%, whereas at $p = 200$, 250 and 300 bar $\tilde{\gamma}$ is reduced by only 24%. It is also worth mentioning, that the addition of cyclohexane has almost no influence on the phase inversion temperature $\tilde{T}$. The experimental errors amount to $\Delta p = \pm 5$ bar, $\Delta \tilde{\gamma} = \pm 0.01$ and $\Delta \tilde{T} = \pm 1$ K.

The studies performed on CO_2-microemulsions, stabilized by Zonyl® or Capstone® surfactants, indicate that the density of CO_2 at the respective temperature and pressure specify the interaction between CO_2 and surfactant molecule and thus the property of the CO_2-microemulsions. Thus, the surfactant mass fractions $\tilde{\gamma}$ for both systems are shown in Figure 3.21 as function of the macroscopic density ρ_{CO_2} of CO_2. The errors for the macroscopic density of CO_2 were estimated assuming $\Delta p = \pm 5$ bar and $\Delta \tilde{\gamma} = \pm 0.01$.

For microemulsions of the type water $- CO_2 -$ Zonyl® FSO 100/FSN 100 it was found, that the surfactant mass fraction $\tilde{\gamma}$ at the $\tilde{X}$-point scales quantitatively with the macroscopic density of CO_2, according to $\ln(\tilde{\gamma}) \sim \rho_{CO_2}$ [30]. Note, that the respective $\tilde{\gamma}(\rho_{CO_2})$-values for the system $H_2O/NaCl - CO_2 -$ Zonyl® FSO 100/FSN 100 ($\delta_{FSN} = 0.40$, $\varepsilon = 0.01$, $p = 150$–300 bar, $\tilde{T}$ = 41.9-35.8 °C) determined by *Klostermann* are also included in Figure 3.21 [30].

As can be seen, the relation $\ln(\tilde{\gamma}) \sim \rho_{CO_2}$ seems to hold true also for the Capstone®-stabilized cyclohexane-free CO_2-microemulsion with $\beta = 0.00$. However, a larger scattering of the data points around the linear trend is found. Furthermore, the $\tilde{\gamma}$–values of the Capstone®-stabilized CO_2-microemulsion are shifted to larger surfactant mass fractions, i.e. Capstone® is slightly less efficient than Zonyl®, which is caused by the fact that the polyfluorinated surfactant tail of the Capstone® surfactant is of the order of one C_2F_4-unit shorter than the Zonyl® surfactant. Comparing the phase inversion temperature $\tilde{T}$ of both systems, lower values of $\tilde{T}$ are found for the Capstone®-stabilized CO_2-microemulsion leading to the conclusion, that also the hydrophilic EO-chain of the Capstone® surfactant is shorter than that of the Zonyl® surfactants.

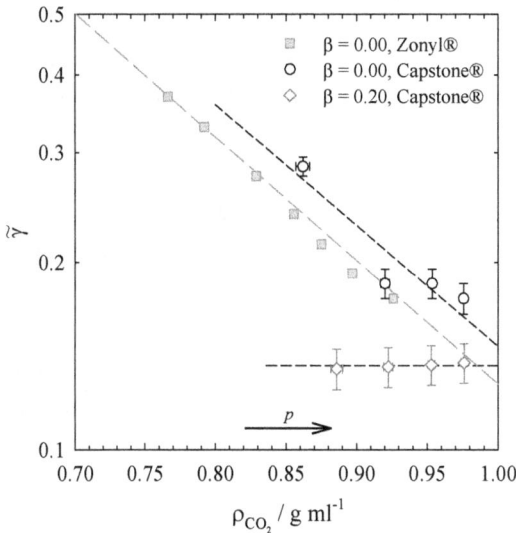

Figure 3.21: Surfactant mass fraction $\tilde{\gamma}$ at the $\tilde{X}$-point of different balanced CO_2-microemulsions plotted as a function of the temperature- and pressure-dependent macroscopic density of CO_2 in a semi-logarithmic plot. Being more precise: $\tilde{\gamma}(\rho_{CO_2})$ of the systems $H_2O/NaClO_4 - CO_2 -$ Capstone® FS-3100 (black circles, $\alpha = 0.40$, $\beta = 0.00$, $\varepsilon = 0.05$, $p = 150$, 200, 250, 300 bar), $H_2O/NaClO_4 - CO_2/$cyclohexane – Capstone® FS-3100 (red diamonds, $\alpha = 0.40$, $\beta = 0.20$, $\varepsilon = 0.05$, $p = 150$, 200, 250, 300 bar) and $H_2O/NaCl - CO_2 -$ Zonyl® FSO 100/FSN 100 (grey squares, $\alpha = 0.40$, $\beta = 0.00$, $\delta_{FSN} = 0.40$, $\varepsilon = 0.01$, $p = 150$, 160, 180, 200, 220, 250, 300 bar). The data of the Zonyl®-containing system were determined by *Klostermann* [30].

Quite contrary, the efficiency of the studied cyclohexane-rich system ($\beta = 0.20$) is found to be independent of the macroscopic density of CO_2. This result might be related to the formation of a concentration gradient within the CO_2-domains, which has been proven to exist by SANS measurements (compare section 2.1.4) [47]. While cyclohexane is enriched in the center of the CO_2-domains, CO_2 molecules are forced to the amphiphilic film, where they increase the local density of CO_2, so that the interactions between CO_2 and the fluorinated surfactant tails are improved. Thus, for the cyclohexane-rich system, the local CO_2-density near the amphiphilic film and not the macroscopic CO_2-density is the determining parameter.

3.2.2 Microstructure of efficiency boosted CO_2-microemulsions

The influence of cyclohexane on the microstructure of the system $H_2O/NaClO_4 -$ CO_2/cyclohexane $-$ Capstone® FS-3100 ($\alpha = 0.40$, $\varepsilon = 0.05$) was studied at $\beta = 0.20$ as a function of pressure at surfactant mass fractions of $\gamma = 0.20$ and 0.30 by means of SANS (Figure 3.22). Again, H_2O was replaced by D_2O in order to increase the scattering length density contrast and decrease the incoherent scattering intensity, while the composition parameters specified are always related to H_2O for a better comparability with the phase diagrams (index p). Note that the replacement of H_2O by D_2O shifts the phase boundaries to lower temperatures by 2–3 K [41]. Furthermore, the only slight variation of $\tilde{T}$ upon pressure increase from $p = 150$ bar to $p = 300$ bar allows to perform the SANS-measurement at a constant temperature of $T = 20.0$ °C at $\gamma_p = 0.20$ and $T = 20.7$ °C at $\gamma_p = 0.30$. Note, that the chosen temperatures are located always in the middle of the one-phase regions of the respective systems. Furthermore, the microstructure of the two-phase ME+L_α sample (compare Figure 3.19) was studied at $\gamma_p = 0.20$, $T = 16.1$ °C $p = 150$ and 300 bar. The recorded scattering curves are shown in Figure 3.22 as a function of the scattering vector q in a double logarithmic plot.

As can be seen, the scattering curves recorded at $\gamma_p = 0.30$ (Figure 3.22, right) show a scattering peak at intermediate q-values, the peak vanishes for the sample prepared at $\gamma_p = 0.20$, indicating a loss of order due to stronger fluctuations at lower surfactant mass fraction. Surprisingly, by increasing the pressure the peak becomes less pronounced and thus indicating a weaker ordering of the structure. However, this effect is due to a superposition of two scattering contributions stemming from the bicontinuous microstructure on the one hand and long range fluctuations on the other hand. The scattering peak and high q-part of the SANS-spectra were analyzed by using the *Teubner-Strey* model (section 2.3.1, eq. 2.39, solid lines) [103] and *Porod* decay for diffuse interfaces (section 2.3.1, eq. 2.41, dashed line) [158, 159], respectively.

The scattering data recorded within the coexistence region ME+L_α at $\gamma_p = 0.20$ and $T = 16.1$ °C show a pronounced scattering peak, which again can be attributed to the scattering contribution of the lamellar phase (compare Figure 3.8). Note that the scattering patterns are anisotropic (see inlet in Figure 3.22, intermediate q-region, $p = 300$ bar), caused by a preferred direction of the domains within the L_α-phase. However, the intensity was radially averaged. The SANS-spectra were described by the combination of two *Teubner-Strey* structure factors (eq. 3.2, short dashed line) [197], revealing similar parameters for the scattering contribution of the bicontinuous phase as gained from the analysis of the one-phase microemulsion recorded at $T = 20.0$ °C.

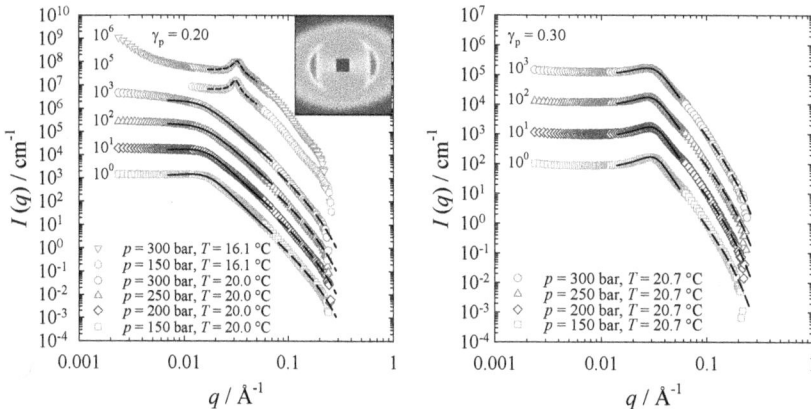

Figure 3.22: Scattering curves of the balanced microemulsion D$_2$O/NaClO$_4$ – CO$_2$/cyclohexane – Capstone® FS-3100 ($\alpha_p = 0.40$, $\beta = 0.20$, $\varepsilon_p = 0.05$) at $\gamma_p = 0.20$ (left) and $\gamma_p = 0.30$ (right). The SANS-curves were recorded at pressures of $p = 150$, 200, 250 and 300 bar, while the temperature was adjusted to $T = 20.0$ and 20.7 °C, respectively. The data were described using the *Teubner-Strey* model [103] (intermediate q-region, solid line) and the *Porod* decay for diffuse interfaces [158, 159] (high q-region, dashed line). In addition for $\gamma - 0.20$ (left) and $p = 150$ and 300 bar, scattering curves in the coexistence region ME+L_α at $T = 16.1$ °C were recorded, exhibiting an anisotropic signal (see inlet for a detector picture at intermediate q-values). The data were described using a combination of two *Teubner-Strey* structure factors [103, 197] (short dashed line).

The structural parameters obtained from the *Teubner-Strey* analysis (intermediate q-region) of the system with $\beta = 0.20$ and $\gamma_p = 0.20$ and 0.30 are listed in Table 3.6.

From the structural parameters a clear trend for each composition is visible. With increasing pressure the characteristic periodicity d_{TS} slightly increase, the correlation length ξ_{TS} decreases and consequently the amphiphilicity factor f_a becomes less negativ. The only exceptions are the SANS-spectra recorded at $\gamma_p = 0.20$ and $p = 250$ and 300 bar, where the *Teubner-Strey* structure factor is not perfectly able to describe the scattering data at low q. As pointed out before, this is presumably due to a superposition of two scattering contributions, while this additional contribution weakens the scattering peak. As a consequence, the amphiphilicity factors $f_a \approx -0.10$ are significantly closer to $f_a = 0$ than the values for the other samples.

Note, that the f_a-trend at $\gamma_p = 0.30$, $p = 150$–300 bar and $\gamma_p = 0.20$, $p \leq 200$ bar is inverse compared to the cyclohexane-free system (section 3.1.2, Table 3.2), where the ordering is slightly weakening with decreasing pressure. Moreover, also ξ_{TS} was found to exhibit an inverse behavior with increasing pressure.

By decreasing the surfactant mass fraction γ (at $p = 150$ bar) from $\gamma_p = 0.30$ to $\gamma_p = 0.20$, the amphiphilicity factor f_a becomes less negative i.e. changes from $f_a = -0.831$ to $f_a = -0.447$, while the characteristic periodicity strongly increases from $d_{TS} = 204$ Å to $d_{TS} = 433$ Å and the correlation length ξ_{TS} increases from $\xi_{TS} = 92.7$ Å to $\xi_{TS} = 111.6$ Å. The f_a-trend was expected, as the peak vanishes and thus the ordering is decreasing with increasing pressure and especially upon reduction of the surfactant mass fraction.

By decreasing the surfactant mass fraction γ in the $\beta = 0.20$ microemulsion at $p = 150$ bar, the amphiphilicity factor f_a decreases strongly from $f_a = -0.782$ at $\gamma_p = 0.30$ to $f_a = -0.447$ at $\gamma_p = 0.20$. Increasing the pressure at $\gamma_p = 0.20$ to $p = 300$ bar the amphiphilicity factor almost reaches zero, i.e. $f_a = -0.095$. However, as discussed before another scattering contribution at low q seems to appear, which causes the disappearance of the peak.

Table 3.6: Structural parameters of the bicontinuous microemulsions $D_2O/NaClO_4$ – CO_2/cyclohexane – Capstone® FS-3100 ($\alpha_p = 0.40$, $\beta = 0.20$, $\varepsilon_p = 0.05$) at surfactant mass fractions of $\gamma_p = 0.20$ and 0.30 and pressures $p = 150$ to 300 bar obtained from the *Teubner-Strey* [103] analysis of the SANS-curves. The analysis of the peak region yields the characteristic periodicity d_{TS}, the correlation length ξ_{TS}, the amphiphilicity factor f_a and the bending rigidities κ_{SANS} and $\kappa_{0,SANS}$.

γ_p	p / bar	T / °C	d_{TS} / Å	ξ_{TS} / Å	f_a	κ_{SANS} / k_BT	$\kappa_{0,SANS}$ / k_BT
0.30	150	20.7	204	92.7	-0.782	0.387	0.918
0.30	200	20.7	205	88.4	-0.760	0.367	0.899
0.30	250	20.7	209	84.0	-0.729	0.342	0.879
0.30	300	20.7	211	76.8	-0.679	0.310	0.849
0.20	150	20.0	433	111.6	-0.447	0.219	0.930
0.20	200	20.0	468	106.4	-0.343	0.193	0.923
0.20	250	20.0	538	94.3	-0.096	0.149	0.912
0.20	300	20.0	551	96.4	-0.095	0.149	0.918

The structural parameters obtained from the analysis of the scattering curves recorded within the ME+L_α-region at $\beta = 0.20$, $\gamma_p = 0.20$ and $T = 16.1$ °C are listed in Table 3.7. The results from the double *Teubner-Strey* analysis (eq. 3.2, short dashed line) of the ME+L_α-sample confirm the expected trends. The amphiphilicity factor f_a of the microemulsion phase was found to be slightly negative, i.e. below the *Lifshitz* line. Thus, one can conclude that again an additional scattering contribution weakens the scattering peak. The amphiphilicity factor of the lamellar phase $f_{a,L_\alpha} = -0.98$ was found to be pressure independent and close to -1, proving again the existence of a L_α phase.

Table 3.7: Structural parameters obtained from the analysis of the SANS-data of the microemulsion $D_2O/NaClO_4$ – CO_2/cyclohexane – Capstone® FS-3100 ($\alpha_p = 0.40$, $\beta = 0.20$, $\gamma_p = 0.20$, $\varepsilon_p = 0.05$) recorded in the coexistence region ME+L_α at $T = 16.1$ °C using the combination of two *Teubner-Strey* structure factors [103, 197]. The analysis yields the characteristic periodicity d_{TS}, the correlation length ξ_{TS} and the amphiphilicity factor f_a of the microemulsion and lamellar phase (index L_α), respectively.

p / bar	d_{TS} / Å	ξ_{TS} / Å	f_a	d_{TS,L_α} / Å	ξ_{TS,L_α} / Å	f_{a,L_α}
150	240	40.5	-0.057	199	346	-0.983
300	210	39.4	-0.162	193	325	-0.982

In order to systematically study the influence of cyclohexane on the microstructure of the Capstone®-stabilized CO_2-microemulsion scattering curves, a cyclohexane content β of $\beta = 0.08$ and $\beta = 0.15$ were also recorded at $\gamma_p = 0.30$. Note, that no complete $T(\gamma)$-section of the SANS-samples were recorded to study the phase behavior of these systems. However, the phase behavior of the SANS-samples is studied adjusting the measuring temperature to the middle of the one-phase region (see section 6.3.2). The microstructure of both systems were studied at 4 pressures between $p = 150$ and 300 bar and $T = 19.0$ °C. The spectra as well as the fits according to the *Teubner-Strey* structure factor (peak-region) [103] and the *Porod* decay for diffuse interfaces (high q-region) [158, 159] are shown in Figure 3.23.

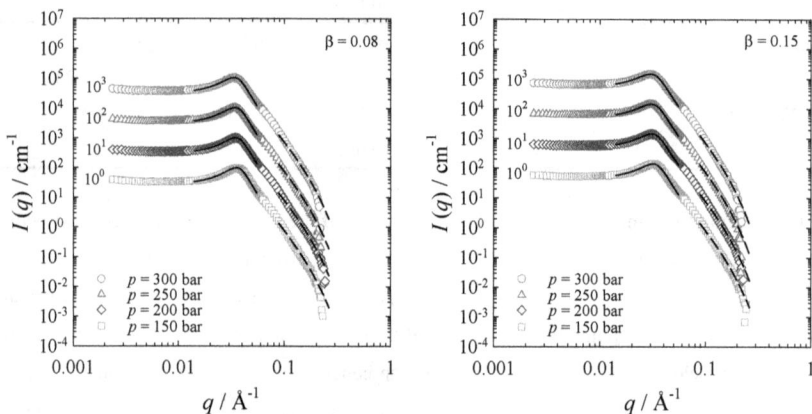

Figure 3.23: Scattering curves of the balanced microemulsion D_2O/$NaClO_4$ – CO_2/cyclohexane – Capstone® FS-3100 ($\alpha_p = 0.40$, $\gamma_p = 0.30$, $\varepsilon_p = 0.05$) at $\beta = 0.08$ (left) and $\beta = 0.15$ (right). The SANS-curves were recorded at pressures of $p = 150$, 200, 250 and 300 bar, the temperature was adjusted to $T = 19.0$ °C for both γ_p-values. The data were described using the *Teubner-Strey* structure factor [103] (peak region, solid line) and the *Porod* decay for diffuse interfaces [158, 159] (high q-region, dashed line).

Comparing the scattering curves of the $\beta = 0.08$ and $\beta = 0.15$ samples with the curves of the $\beta = 0.20$ sample (see Figure 3.22, right), one clearly recognizes that the peak becomes more and more pronounced with decreasing cyclohexane-content β. However, compared to the cyclohexane-free CO_2-microemulsion a more pronounced scattering peak is found. A more quantitative picture with respect to the influence of cyclohexane on the microstructure is obtained if one compares the structural parameters gained from the analysis of the scattering peaks, which are listed in Table 3.6 and Table 3.8.

As obtained for the cyclohexane-free system, the characteristic periodicity d_{TS} of the bicontinuous structure increases slightly with increasing pressure due to the increasing solubility of surfactant in CO_2. Because at the same time the correlation length ξ_{TS} is almost constant for $\beta = 0.08$ but decreases for $\beta = 0.15$ and $\beta = 0.20$, also the amphiphilicity factor f_a become less negative, indicating a weakening of the structure. Note, that the f_a-trend of the cyclohexane-free system (section 3.1.2, Table 3.2) is inverse, i.e. a slight ordering of the structure is found with increasing pressure. This trend becomes stronger with increasing β. At a constant pressure of $p = 150$ bar f_a runs from $f_a = -0.792$ for $\beta = 0.00$ over $f_a = -0.856$ for $\beta = 0.08$ over $f_a = -0.831$ for $\beta = 0.15$ to $f_a = -0.782$ for $\beta = 0.20$. Thus, the amphiphilicity factor passes a minimum at $\beta \approx 0.08$ and a larger efficiency boosting is expected for this cyclohexane mass fraction. Consequently the largest values of the renormalized bending rigidity are found for the CO_2-microemulsion containing cyclohexane with a mass fraction of $\beta = 0.08$.

Table 3.8: Structural parameters of the bicontinuous microemulsions $D_2O/NaClO_4$ – CO_2/cyclohexane – Capstone® FS-3100 ($\alpha_p = 0.40$, $\gamma_p = 0.30$, $\varepsilon_p = 0.05$) at cyclohexane mass fractions of $\beta = 0.08$ and 0.15 and $p = 150$–300 bar obtained from the *Teubner-Strey* [103] analysis of the SANS-curves. The analysis of the peak region yields the characteristic periodicity d_{TS}, the correlation length ξ_{TS}, the amphiphilicity factor f_a and the bending rigidities κ_{SANS} and $\kappa_{0,SANS}$.

β	p / bar	T / °C	d_{TS} / Å	ξ_{TS} / Å	f_a	κ_{SANS} / k_BT	$\kappa_{0,SANS}$ / k_BT
0.08	150	19.0	173	98.9	-0.856	0.485	0.978
0.08	200	19.0	176	99.1	-0.853	0.479	0.976
0.08	250	19.0	178	99.3	-0.849	0.474	0.973
0.08	300	19.0	182	99.8	-0.845	0.467	0.971
0.15	150	19.0	191	100.2	-0.831	0.445	0.962
0.15	200	19.0	195	99.4	-0.822	0.433	0.954
0.15	250	19.0	197	98.8	-0.817	0.426	0.949
0.15	300	19.0	200	97.3	-0.807	0.414	0.940

Moreover, from the analysis of the large q-part with the *Porod* decay for diffuse interfaces the specific internal interface S/V is obtained using the relation $\phi_{C,i} = l_C \cdot S/V$ [56, 158]. The interfacial volume fraction of surfactant $\phi_{C,i}$ can be calculated estimating the length of the surfactant as $l_C = 11$ Å. As can be seen in Table 3.9, $\phi_{C,i}$ decreases slightly with increasing pressure and increasing β. This explains the slight increase of the characteristic periodicity d_{TS} as the overall size of the amphiphilic film shrinks. To highlight the trend of the characteristic periodicity d_{TS} as function of the cyclohexane mass fractions β, the reduced characteristic length scale $\xi \cdot \phi_{C,i}$ with $\xi = d_{TS}/2$ was considered in the next step. For microemulsions stabilized by the same surfactant, a constant value is expected for the reduced characteristic length scale $\xi \cdot \phi_{C,i} = 6 \cdot l_C \cdot \phi(1-\phi)$ (eq. 3.4). However, as can be seen in Figure 3.24, the value of $\xi \cdot \phi_{C,i}$ slightly increases linearly with increasing cyclohexane-content β for both surfactant mass fractions $\gamma_p = 0.20$ (left) and $\gamma_p = 0.30$ (right). This trend is a consequence of the term $\phi(1-\phi)$ or the surfactant length l_C. Indeed at $\beta = 0.00$ with increasing pressure $\phi(1-\phi)$ slightly decreases from $\phi(1-\phi) = 0.243$ for $p = 150$ bar to $\phi(1-\phi) = 0.239$ for $p = 300$ bar. Moreover, at a constant pressure $p = 150$ bar, $\phi(1-\phi)$ increases slightly with increasing cyclohexane mass fraction from $\phi(1-\phi) = 0.243$ for $\beta = 0.00$ to $\phi(1-\phi) = 0.249$ for $\beta = 0.20$. Thus, both observations, the slight increase of $\xi \cdot \phi_{C,i}$ with increasing cyclohexane mass fraction β and with decreasing pressure, can be explained by the pressure-dependent $\phi(1-\phi)$-term. In addition, also the surfactant length l_C, which was assumed to be $l_C = 11$ Å, might contribute to the observed trends.

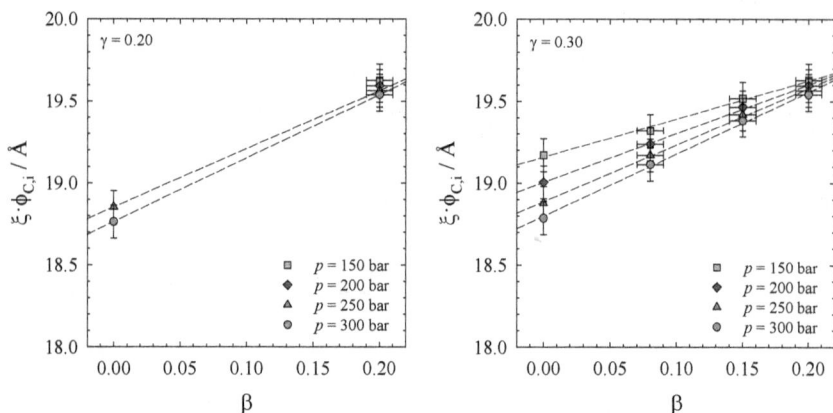

Figure 3.24: Scaled characteristic length scales $\xi \cdot \phi_{C,i}$ of the system $D_2O/NaClO_4 -$ CO_2/cyclohexane – Capstone® FS-3100 ($\alpha_p = 0.40$, $\varepsilon_p = 0.05$) as function of the mass fraction of cyclohexane in the CO_2/cyclohexane mixture β at $\gamma_p = 0.20$ (left) and $\gamma_p = 0.30$ (right) for $p = 150$, 200, 250 and 300 bar. Note, that at $\gamma_p = 0.20$ and $\beta = 0.00$ SANS-measurements were only performed at $p = 250$ and 300 bar.

The volume fraction of monomerically solubilized surfactant molecules $\phi_{C,mon,b}$ in CO_2 was found to be increasing as expected with increasing pressure as well as slightly increase with increasing cyclohexane mass fraction from $\phi_{C,mon,b} = 0.081$ ($\beta = 0.00$) to $\phi_{C,mon,b} = 0.097$ at $\beta = 0.20$ for $p = 300$ bar and $\gamma_p = 0.30$ (see 3.9).

Table 3.9: Scattering length density contrast $\Delta\rho_s$, specific internal interface S/V, diffusivity of the amphiphilic film t, volume fraction of surfactant ϕ_C, interfacial volume fraction of surfactant $\phi_{C,i}$ and volume fraction of surfactant monomerically solubilized in CO_2 $\phi_{C,mon,b}$ resulting from the *Porod*-analysis [158, 159] of the high q-range parts of scattering spectra of the balanced microemulsion $D_2O/NaClO_4 - CO_2/cyclohexane - Capstone®$ FS-3100 ($\alpha_p = 0.40$, $\varepsilon_p = 0.05$) at different surfactant mass fractions γ_p, cyclohexane mass fractions β and pressures p.

γ_p	β	p / bar	T / °C	$\Delta\rho_s$ / 10^{-6} Å^{-2}	S/V / Å^{-1}	t / Å	ϕ_C	$\phi_{C,i}$	$\phi_{C,mon,b}$
0.30	0.08	150	19.0	4.2	0.0203	6.0	0.230	0.223	0.020
0.30	0.08	200	19.0	4.2	0.0199	6.0	0.233	0.219	0.042
0.30	0.08	250	19.0	4.2	0.0196	6.0	0.235	0.216	0.057
0.30	0.08	300	19.0	4.1	0.0191	6.0	0.237	0.210	0.078
0.30	0.15	150	19.0	4.3	0.0186	6.2	0.223	0.205	0.050
0.30	0.15	200	19.0	4.2	0.0182	6.2	0.226	0.200	0.069
0.30	0.15	250	19.0	4.1	0.0179	6.6	0.228	0.197	0.084
0.30	0.15	300	19.0	4.2	0.1760	6.6	0.230	0.194	0.097
0.30	0.20	150	20.7	4.3	0.0175	6.8	0.217	0.193	0.062
0.30	0.20	200	20.7	4.3	0.0174	6.6	0.220	0.191	0.073
0.30	0.20	250	20.7	4.3	0.0170	6.4	0.222	0.187	0.089
0.30	0.20	300	20.7	4.3	0.0168	6.2	0.223	0.185	0.097
0.20	0.20	150	20.0	4.3	0.0082	5.8	0.139	0.091	0.107
0.20	0.20	200	20.0	4.3	0.0076	5.5	0.141	0.084	0.125
0.20	0.20	250	20.0	4.2	0.0066	5.5	0.142	0.073	0.149
0.20	0.20	300	20.0	4.2	0.0064	5.5	0.144	0.071	0.158

The bending rigidity κ_{SANS}/k_BT and renormalized corrected bending rigidity $\kappa_{0,SANS}/k_BT$ obtained from the analysis applying the model of *Safran* and *Pieruschka* [156] for the various compositions of the microemulsion system $D_2O/NaClO_4 - CO_2/cyclohexane - Capstone®$ FS-3100 (listed in Table 3.9) are plotted in Figure 3.25 as function of the pressure.

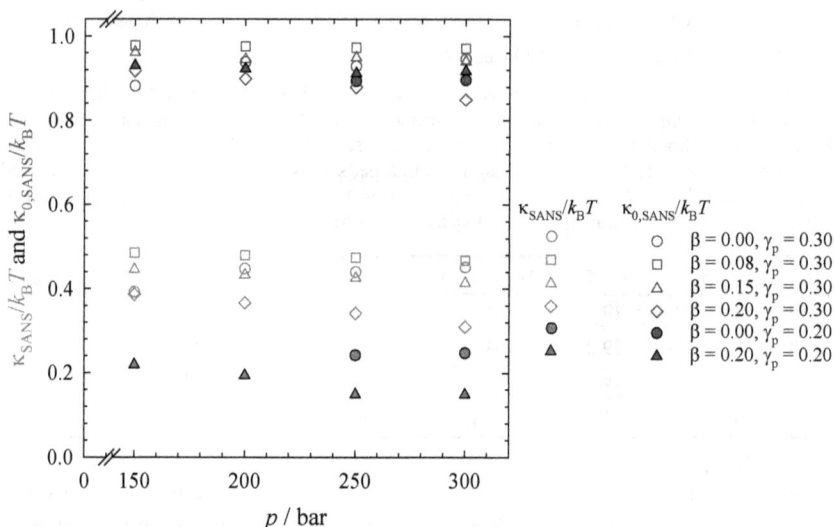

Figure 3.25: Bending rigidity κ_{SANS}/k_BT (red symbols) and renormalized corrected bending rigidity $\kappa_{0,SANS}/k_BT$ (blue symbols) obtained for the amphiphilic film of the microemulsion $D_2O/NaClO_4 - CO_2/cyclohexane - Capstone®$ FS-3100 ($\alpha_p = 0.40$, $\varepsilon_p = 0.05$) at $\gamma_p = 0.20$ (filled symbols) and $\gamma_p = 0.30$ (hollow symbols) for cyclohexane-contents of $\beta = 0$, 0.08, 0.15 and 0.20 as function of the pressure p. Note that the values were obtained using the model of *Safran* and *Pieruschka* [156] (eq. 3.7 and 3.8).

As can be seen in Figure 3.25, with $\gamma_p = 0.30$ the largest values for κ_{SANS} and $\kappa_{0,SANS}$ are found for the CO_2-microemulsion with $\beta = 0.08$. Smaller but similar bending rigidities are observed for the CO_2-microemulsion with $\beta = 0.00$ and $\beta = 0.15$, while the smallest bending rigidities are found for the $\beta = 0.20$ CO_2-microemulsion $\gamma_p = 0.20$ and 0.30. Note that almost constant values of the rigidity with increasing pressure are found for the respective composition of the sample. Only for the system at $\beta = 0.20$ κ_{SANS} and $\kappa_{0,SANS}$ decreases slightly for both γ_p-values.

3.2.3 Dynamics of bicontinuous CO_2-microemulsions containing the co-oil cyclohexane

In the two sections before (sections 3.2.1 and 3.2.2), it was shown, that the surfactant mass fraction $\tilde{\gamma}$ at the optimum point $\tilde{X}$ can be decreased upon the partial replacement of CO_2 by the co-oil cyclohexane. Moreover, the ordering of the structure is found to run through a maximum around the sample with $\beta = 0.08$ indicated by a minimum (most negative) amphiphilicity factor f_a and the largest bending rigidities κ_{SANS} and $\kappa_{0,SANS}$. In order to elucidate, whether these values are reliable, NSE-studies were performed. As pointed out in section 3.1.4 the analysis of the NSE-data can be performed on different levels. The most systematic results which are in agreement with literature values for similar CO_2-microemulsions were obtained using the linear combination of two independent relaxation rates and the non-numerical solution for the *Zilman-Granek* relaxation rate (eq. 3.13). Thus, the same analysis was used to analyze the NSE-data of the cyclohexane-containing CO_2-microemulsion systems. Solving eq. 3.10, the bending rigidities κ_{NSE}/k_BT for the system $D_2O/NaClO_4 - CO_2$/cyclohexane –Capstone® FS-3100 ($\alpha_p = 0.40$, $\varepsilon_p = 0.05$) at $\beta = 0.08$ and 0.15 ($\gamma_p = 0.30$) and at $\beta = 0.20$ ($\gamma_p = 0.30$ and 0.20) were determined and shown in Figure 3.26 and Figure 3.27, respectively. In addition, the values of the renormalization corrected bending rigidity from SANS $\kappa_{0,SANS}$ (short dashed line) is given.

As for the cyclohexane-free microemulsion the values for the bending rigidity κ_{NSE} determined by NSE are slightly lower than the $\kappa_{0,SANS}$-values obtained via SANS. The κ_{NSE}-values of the system at $\beta = 0.20$ (Figure 3.26) were found to slightly increase with increasing pressure for $\gamma_p = 0.20$ and 0.30. Increasing the surfactant mass fraction γ at constant β-values, the κ_{NSE}-values are found to significantly decrease by $\Delta\kappa_{NSE}/k_BT \approx 0.1$. This trend is consistent with the bending rigidities from SANS.

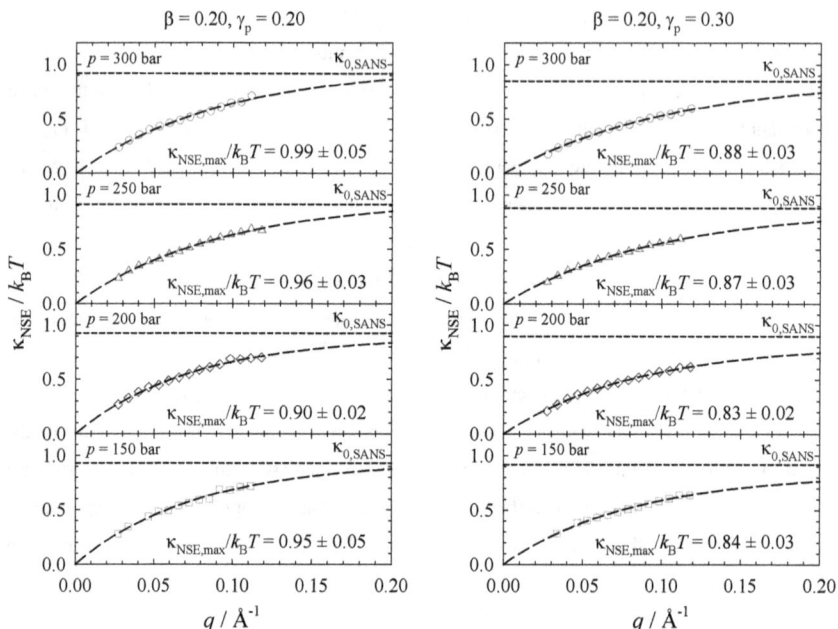

Figure 3.26: Bending rigidity $\kappa_{NSE}/k_B T$ of the amphiphilic film in the bicontinuous microemulsion system $D_2O/NaClO_4 - CO_2/cyclohexane - Capstone® FS-3100$ ($\alpha_p = 0.40$, $\beta = 0.20$, $\varepsilon_p = 0.05$, $p = 150$, 200, 250, 300 bar) at $\gamma_p = 0.20$, $T = 20.7\,°C$ (left) and $\gamma_p = 0.30$, $T = 20.0\,°C$ (right). The bending rigidities were determined from evaluation of the NSE-data with $S(q,\tau)/S(q,0) = A_D \cdot \exp(-\Gamma_D \cdot \tau) + (1-A_D) \cdot \exp(-(\Gamma_{ZG} \cdot \tau)^{2/3})$ and the solution of the *Zilman-Granek* relaxation rate Γ_{ZG} as function of the scattering vector q. The bending rigidity rises exponential to a maximum value $\kappa_{NSE,max}/k_B T$ (long dashed lines). For comparison the respective renormalized corrected bending rigidities obtained from SANS $\kappa_{0,SANS}$ are shown (short dashed line).

To systematically study the influence of cyclohexane on the bending rigidity, additional NSE-measurements at $\beta = 0.08$ and 0.15 ($\gamma_p = 0.30$) and pressures of $p = 150$ and 300 bar were performed (Figure 3.27). Here, a slight decrease of the bending rigidity with increasing pressure from $p = 150$ bar to $p = 300$ bar was found, while a higher cyclohexane concentration β only marginally increases the bending rigidity of the amphiphilic film.

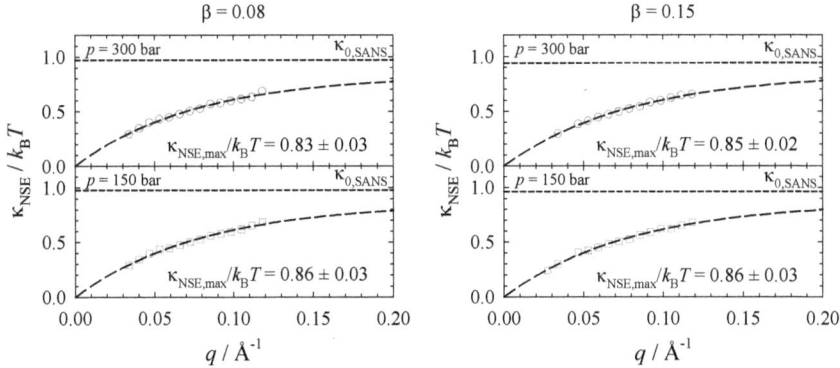

Figure 3.27: Bending rigidity κ_{NSE}/k_BT of the amphiphilic film in the bicontinuous microemulsion system $D_2O/NaClO_4 - CO_2/cyclohexane - Capstone® FS-3100$ ($\alpha_p = 0.40$, $\gamma_p = 0.30$, $\varepsilon_p = 0.05$, $p = 150$, 300 bar, $T = 19.0$ °C) at $\beta = 0.08$ (left) and $\beta = 0.15$ (right). The bending rigidity were determined from evaluation of the NSE-data with $S(q,\tau)/S(q,0) = A_D \cdot exp(-\Gamma_D \cdot \tau) + (1-A_D) \cdot exp(-(\Gamma_{ZG} \cdot \tau)^{2/3})$ and the solution of the *Zilman-Granek* relaxation rate Γ_{ZG} as function of the scattering vector q. The bending rigidity rises exponential to a maximum value $\kappa_{NSE,max}/k_BT$ (long dashed lines). For comparison the respective renormalized corrected bending rigidities obtained from SANS $\kappa_{0,SANS}$ are shown (short dashed line).

Please note, that the p- and β-dependent κ_{NSE}-values at $\gamma_p = 0.30$ are quite similar and no clear trend can be observed. Comparing $\kappa_{0,SANS}$ and κ_{NSE}-values, for the $\beta = 0.08$ and $\beta = 0.15$ system, considerable smaller values are found for κ_{NSE}.

In conclusion, while the $\kappa_{0,SANS}$-values indicate that the most efficient and structured CO_2-microemulsion can be formulated adjusting $\beta = 0.08$, the largest κ_{NSE}-values are found for $\beta = 0.08$, too. However, please note that the analysis of the NSE-data was not obviously because of the small differences of the diffusion and fluctuation rates, as both processes occur at the same time scale.

4 Polyol-rich microemulsion systems

The prospect of generating nanocellular PU foams via the "Principle Of Supercritical Microemulsion Expansion" (POSME) was the main driving force to formulate and study the properties of CO_2-microemulsions. In the POSME concept microemulsions consisting of nano-pools of supercritical CO_2 in a polymerizable material (here polyol) are expanded adding isocyanate shortly beforehand. Adjusting supercritical conditions the nucleation of gaseous CO_2 is avoided so that each CO_2-swollen micelle grows to a nanopore in the PU foam. However, the results obtained within the last years have shown that the high monomeric solubility of CO_2 in polyol leads to the nucleation of ill-defined CO_2 bubbles in the continuous polyol matrix. Thereby the nucleation of a gaseous phase in polyol dominates and instead of nanocellular microcellular pores are formed. Furthermore, the high monomeric solubility of CO_2 in polyol favors the *Ostwald* ripening of the CO_2 nano-pools. Thus the formulation and characterization of appropriate, foamable CO_2-in-polyol microemulsions is the first task towards the generation of nanocellular PU foams (section 4.1). Note that the experimental methods are described in the appendix (section 6.3).

Having formulated CO_2-in-polyol microemulsions, the next step was to proof the existence of CO_2 nano-pools in a polyol matrix (section 4.2) applying small angle neutron scattering (SANS). Using a partly-deuterated polyol not only the existence of micelles in the binary polyol – non-ionic surfactant system but also the swelling of these structures by CO_2 could be proven. Finally, first studies with respect to the change of the microstructure upon expansion were performed conducting periodic pressure jumps in combination with SANS.

Following an idea of *Strey et al.* [39] in section 4.3 the blowing agent CO_2 was partly replaced by a low-molecular, hydrophobic co-oil, which was proposed to slow down the expansion-induced coarsening of the structure, i.e. acting as an anti-aging agent (AAA). Therefore the influence of different co-oils on the phase behavior, microstructure and kinetics of foamable CO_2-in-polyol microemulsion systems was determined.

4.1 Formulation of a polyol-rich CO_2-microemulsion

First microemulsions of the type polyol – alkane – surfactant were formulated in 2006 at ambient pressure within the cooperation "nanoPUR" between Prof. *Strey* and the Bayer

MaterialScience AG (BMS), which started in 2005 (see Introduction). Then the alkanes were replaced by CO_2 at $p > p_c$. Note, that at first environmental unfriendly fluorinated surfactants were used [109], since they are known to solubilize CO_2 in an efficient way (see first part of this thesis, section 3) [42, 43]. Furthermore, polyol-microemulsions with supercritical propane as well as 1,1,1,2-tetrafluorethane as hydrophobic blowing agent were formulated [37, 212]. While the better interaction between these fluids and the surfactant molecules is advantageous with regard to the POSME process, the use of propane is disadvantageous in view of safety concerns. Moreover, the global warming potential of both gases are higher than the one of carbon dioxide. The use of CO_2 thus implies a lower environmental impact. In 2009 the first polyol – CO_2 – non-ionic, non-fluorinated surfactant microemulsion system was formulated using the polyethersiloxane surfactant Dow Corning® Q2-5211 [109].

Trisiloxane surfactants (M(D'E$_j$)M), e.g. the used Dow Corning® Q2-5211 (molecular structure is shown in Figure 4.1, chemically equivalent to Xiameter^{TM} OFX-5211), have a good affinity towards CO_2 due to the stubby, i.e. short and bulky, CO_2-philic tail group [121, 213, 214]. M and D' are the short forms of $Si(CH_3)_3O_{1/2}$- and $-O_{1/2}Si(CH_3)(R)O_{1/2}$-, respectively. Note, that R is a -$(CH_2)_3O$- spacer between the siloxane and the hydrophilic ethoxy units E [215].

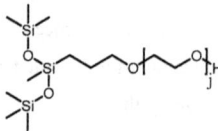

Figure 4.1: Molecular structure of trisiloxane surfactants (M(D'E$_j$)M). In case of Q2-5211, the average number of ethoxy units is $j \approx 10$.

Polyether-polysiloxanes with larger molecular mass are commonly used as foam stabilizers in PU foams.

The polyol-rich component used for the formulation of the first polyol – CO_2 – non-fluorinated surfactant microemulsion system consisted of the bifunctional polyester polyol Desmophen® VP.PU 1431, glycerol and tris(2-chloroisopropyl) phosphate (TCPP) [37, 109]. Thereby, the amount of the three components in the polyol-rich component is defined by the parameter

$$\psi_{Ai} = \frac{m_{Ai}}{\sum_i m_{Ai}} \, . \qquad\qquad 4.1$$

In this formulation, 13.6 wt% of glycerol ($\psi_{glycerol} = 0.136$) and TCPP ($\psi_{TCPP} = 0.136$) were added in order to increase the functionality of the polyol and the ignition temperature of the foam, respectively [3, 216]. Furthermore, glycerol is more polar than Desmophen® VP.PU 1431. Thus upon addition of glycerol, the miscibility gap of the binary polyol – CO_2 system is expected to increase, equivalent to a reduction of the monomeric solubility of CO_2 in the polyol.

In recent years, the influence of different (co-)surfactants, co-oils or hydrophilic components and further parameters on the phase behavior of polyol-rich CO_2-microemulsions were studied [37, 38, 212, 217]. With respect to optimal foam properties, the CO_2 mass fraction α in the polyol-CO_2 mixture was varied between $\alpha = 0.10$ and $\alpha = 0.30$ [37]. While too low mass fractions α yield foams with high density and thick pore walls, too large α-values lead to a tearing of the foam.

An unexpected feature of polyol-containing CO_2-microemulsions is the phase sequence $\overline{2} \rightarrow 3 \rightarrow \underline{2}$ (for low and intermediate surfactant mass fractions γ), respectively, $\overline{2} \rightarrow 1 \rightarrow \underline{2}$ (for large γ) observed with increasing temperature. This phase sequence is inverse compared to classical microemulsions of the type water – oil – non-ionic surfactant (see section 2.1.1). In order to elucidate the cause of the unexpected phase sequence the respective binary side systems were recently studied in detail by *Tseng* [218]. In Figure 4.2 the phase diagrams of the binary systems VP.PU 1431/glycerol/TCPP – Q2-5211 ($p = 1$ bar, $\psi_{1431} = 0.728$, $\psi_{glycerol} = 0.136$, $\psi_{TCPP} = 0.136$) and CO_2 – Q2-5211 at $p = 120$ bar and $p = 160$ bar are shown [218]. In addition the phase diagrams of the corresponding ternary system VP.PU 1431/glycerol/TCPP – CO_2 – Q2-5211 ($\psi_{1431} = 0.728$, $\psi_{glycerol} = 0.136$ and $\psi_{TCPP} = 0.136$) at $\alpha = 0.15$ and $p = 120$ and 160 bar recorded in this work are displayed (Figure 4.2, right).

As can be seen, only a lower miscibility gap was found for the system VP.PU 1431/glycerol/TCPP – Q2-5211 ($\psi_{1431} = 0.728$, $\psi_{glycerol} = 0.136$, $\psi_{TCPP} = 0.136$), while the system H_2O – non-ionic surfactant exhibits both a lower (typically below 0 °C) and an upper miscibility gap. Hence, the non-ionic trisiloxane surfactant becomes increasingly soluble in polyol with increasing temperature. Please note, that due to the low compressibility of polyol and Q2-5211 the pressure dependence of the phase boundary (typically 3.5 K per 100 bar [108]) is weak. The phase diagram of the binary system CO_2 – Q2-5211 shown for $p = 120$ and 160 bar exhibits an upper miscibility gap, which is caused by the decreasing CO_2 density and thus the decreasing CO_2-surfactant interaction, leading to a demixing with increasing temperature. Consequently, by raising the pressure the density of CO_2 is increased and the miscibility gap shrinks considerably [218]. Thus, in general the phase behavior is inverse compared to oil – non-ionic surfactant systems, which show a lower miscibility gap.

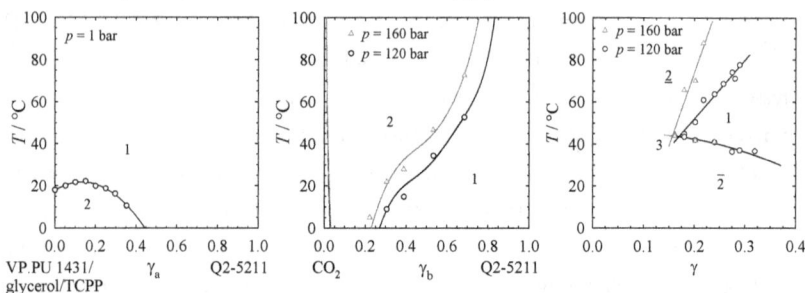

Figure 4.2: *Left*: Phase diagram of the pseudo-binary system VP.PU 1431/glycerol/TCPP – Q2-5211 ($\psi_{1431} = 0.728$, $\psi_{glycerol} = 0.136$, $\psi_{TCPP} = 0.136$) at $p = 1$ bar as function of temperature T and surfactant mass fraction γ_a. The system exhibits only a lower miscibility gap, while the system H_2O – non-ionic surfactant shows also an upper miscibility gap. *Middle*: Phase diagram of the CO_2 – Q2-5211 system at $p = 120$ bar (black circles) and $p = 160$ bar (red triangles). An upper miscibility gap is found, i.e. the phase behavior is inverse compared to systems of the type oil – non-ionic surfactant. *Right*: $T(\gamma)$-sections of the ternary system VP.PU 1431/glycerol/ TCPP – CO_2 – Q2-5211 ($\psi_{1431} = 0.728$, $\psi_{glycerol} = 0.136$, $\psi_{TCPP} = 0.136$) at $\alpha = 0.15$ and $p = 120$ bar (black circles) and $p = 160$ bar (red triangles) exhibiting the phase sequence $\overline{2} \to 3 \to \underline{2}$ or $\overline{2} \to 1 \to \underline{2}$ as a consequence of the binary miscibility gaps. The phase diagrams of the binary systems are taken from *Tseng* [218].

As predicted from the phase diagrams of the binary side systems, the phase diagrams of the ternary system VP.PU 1431/glycerol/TCPP – CO_2 – Q2-5211 ($\alpha = 0.15$, $p = 120$ and 160 bar, $\psi_{1431} = 0.728$, $\psi_{glycerol} = 0.136$, $\psi_{TCPP} = 0.136$) which are shown in Figure 4.2, right, exhibit the phase sequence $\overline{2} \to 1 \to \underline{2}$ with increasing temperature. Hence, at low temperatures a surfactant-rich microemulsion phase coexists with a polyol-excess phase (denoted as $\overline{2}$). Thereby the lower phase boundary $\overline{2} \to 1$ results from the almost pressure-independent polyol – Q2-5211 miscibility gap. Note, that in both phases a considerable amount of CO_2 is monomerically solubilized in the polyol. Speculating about the structure of the lower surfactant-rich microemulsion phase, we propose the existence of CO_2-swollen micelles in a continuous polyol matrix. SANS-measurements will be used to prove this suggestion. At intermediate temperatures a one-phase region (1) can be found, while at high temperatures a polyol-rich phase containing CO_2-swollen micelles coexists with a CO_2-excess phase (denoted as $\underline{2}$). As the upper phase boundary $1 \to \underline{2}$ is a consequence of the miscibility gap of the binary CO_2 – Q2-5211 system, it is found to be highly pressure-dependent. Furthermore, a three-phase region (3) can be found at low surfactant mass fractions and intermediate temperatures as a consequence of the superposition of the binary miscibility gaps.

4.1.1 Surface tension

In order to prove, whether surfactant micelles are formed in the polyol-rich component, surface tension measurements were performed. Furthermore, the monomeric solubility of the trisiloxane surfactant in the used polyols can be determined directly from the critical micelle concentration (*cmc*). In more detail, the surface tension of Q2-5211 in water and VP.PU 1431/glycerol/TCPP ($\psi_{1431} = 0.728$, $\psi_{glycerol} = 0.136$, $\psi_{TCPP} = 0.136$) was determined as function of the surfactant concentration ϕ_C using the *Du Nuoy* ring method (see section 6.3.5) [88, 219].

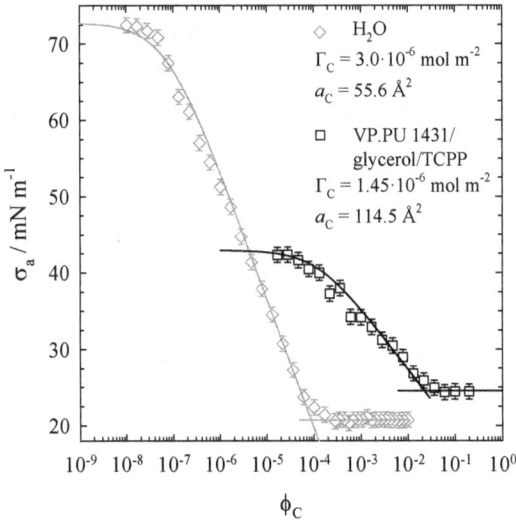

Figure 4.3: Semi-logarithmic plot of the surface tension σ_a as function of the volume fraction ϕ_C of the systems $H_2O - Q2$-5211 (grey diamonds) and polyol – Q2-5211 (black squares) at $p = 1$ bar and $T = 21 \pm 1$ °C. The polyol mixture consists of VP.PU 1431 ($\psi_{1431} = 0.728$), glycerol ($\psi_{glycerol} = 0.136$) and TCPP ($\psi_{TCPP} = 0.136$). Replacing H_2O by the polyol mixture the *cmc* increases by more than 2 orders of magnitude from $\phi_{C,mon,a} = 8.7 \cdot 10^{-5}$ to $\phi_{C,mon,a} = 2.2 \cdot 10^{-2}$, respectively. Simultaneously, the surface tension at the *cmc* increases slightly from 20.7 mN·m^{-1} for H_2O to 24.5 mN·m^{-1} for VP.PU 1431/glycerol/TCPP. [88]

As can be seen in Figure 4.3, the surface tension σ_a decreases for both systems on addition of surfactant starting at $\sigma_0 = 72.0$ mN·m^{-1} for water and $\sigma_0 = 42.9$ mN·m^{-1} in case of

VP.PU 1431/glycerol/TCPP (ψ_{1431} = 0.728, $\psi_{glycerol}$ = 0.136, ψ_{TCPP} = 0.136). The latter value is comparable to polyethylene oxide (σ_0 = 42.9 mN·m^{-1} at T = 20 °C) [220]. Upon further addition of Q2-5211 the surface tensions eventually reaches a constant value of σ_{cmc} = 20.7 mN·m^{-1} at $\phi_{C,mon,a}$ = 8.7·10^{-5} for H$_2$O and of σ_{cmc} = 24.5 mN·m^{-1} at $\phi_{C,mon,a}$ = 2.2·10^{-2} for VP.PU 1431/glycerol/TCPP, respectively. Thus, the *cmc* of Q2-5211 in the polyol-rich component is a factor of 250 larger than in H$_2$O. However, these results confirm, that surfactant micelles are formed in the polyol-rich component. In addition to the increase of the *cmc*, the area per molecule a_C obtained from the analysis of the data by the *Gibbs* adsorption isotherme (eq. 2.9) is doubling from a_C = 55.6 Å^2 to a_C = 114.5 Å^2. Note, that the values found for the interfacial concentration Γ_C, a_C, σ_{cmc} and the *cmc* of the system H$_2$O – Q2-5211 are similar to ones for H$_2$O – M(D'E$_j$)M with j = 8 and 12 described in literature [221].

The much larger *cmc*- and Γ_C-values are caused by the fact, that polyol molecules are less polar and larger than water molecules. Thus, the hydrophobic effect, i.e. the unfavorable interactions between the hydrophobic siloxane part of the surfactant and polyol, is less pronounced and therewith explains the increase of *cmc*, σ_{cmc} and Γ_C. For a more detailed study of the influence of the polyol-composition on the surface tension, *cmc* and the surface area per molecule see [88].

4.1.2 Phase behavior

Having in mind the technical implementation of the POSME process for the production of nanocellular polyurethane foams, the requirements for the polyol-rich CO$_2$-microemulsions are decisive. Due to possible pressure and temperature gradients as well as CO$_2$-concentration gradients within the pilot plant the phase behavior of the polyol-rich CO$_2$-microemulsions had to be studied as function of pressure, temperature and the mass fraction α of CO$_2$ beforehand. Thus, the temperature-dependent phase behavior of the system VP.PU 1431/glycerol/TCPP – CO$_2$ – Q2-5211 (ψ_{1431} = 0.728, $\psi_{glycerol}$ = 0.136, ψ_{TCPP} = 0.136) was systematically studied between α = 0.10–0.20 and p = 80–200 bar. The influence of the pressure on the phase behavior of this system is shown in Figure 4.4 at α = 0.15. To account for the strong decrease of the CO$_2$ density below p = 120 bar, the phase behavior was studied at six pressures p = 80, 90, 100, 120, 160 and 200 bar.

As can be seen, with increasing temperature the inverse phase sequence $\overline{2} \rightarrow 1 \rightarrow \underline{2}$ is found in general. The efficiency of the surfactant Q2-5211 to solubilize CO$_2$ in the polyol mixture strongly decreases with decreasing pressure. Thus the optimum point $\widetilde{X}$ shifts from $\widetilde{\gamma}$ = 0.136 at

$p = 200$ bar to $\tilde{\gamma} = 0.287$ at $p = 80$ bar. This trend can be deduced from the phase behavior of the binary side systems. As pointed out before, the phase behavior of the binary polyol – Q2-5211 system is almost pressure-independent. Consequently, the lower phase boundary $\overline{2} \rightarrow 1$ is almost pressure-independent. The upper phase boundary $1 \rightarrow \underline{2}$, which is caused by the decreasing miscibility gap of the system CO_2 – Q2-5211, however, shifts to smaller values of γ. In addition, the monomeric solubility of CO_2 in the polyol mixture increases with increasing pressure and therewith contributing to the shift of $\tilde{\gamma}$ to lower values in efficiency (shown in more detail in Figure 4.8). Hence, less CO_2 has to be solubilized by the surfactant.

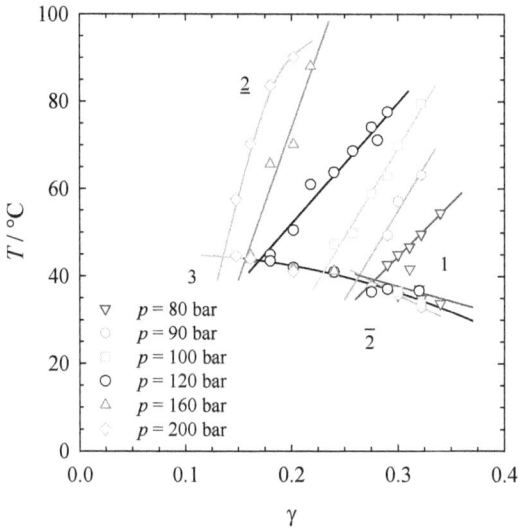

Figure 4.4: $T(\gamma)$-sections for the system VP.PU 1431/glycerol/TCPP – CO_2 – Q2-5211 ($\psi_{1431} = 0.728$, $\psi_{\text{glycerol}} = 0.136$, $\psi_{\text{TCPP}} = 0.136$) at $\alpha = 0.15$ and 6 different pressures between $p = 80$ and 200 bar. With increasing pressure the efficiency of the surfactant Q2-5211 to solubilize CO_2 in the polyol mixture increases. Please note, that the lower phase boundary is almost pressure-independent.

In the next step the influence of the amount of CO_2 on the phase behavior of the system was studied determining the phase diagram at $\alpha = 0.10$ and $\alpha = 0.20$ for 3 different pressure values between $p = 120$ and 200 bar (Figure 4.5 and Figure 4.6, respectively).

The decrease of the mass fraction of CO_2 in the polyol/CO_2-mixture to $\alpha = 0.10$ drastically shifts the phase boundaries to lower values of γ. E.g., the surfactant mass fraction required for the formulation of a one-phase mixture at $\alpha = 0.10$ decreases from $\tilde{\gamma} = 0.056$ at $p = 120$ bar to $\tilde{\gamma}$ $= 0.005$ at $p = 160$ bar. At $p = 200$ bar even a one-phase mixture was observed between $T = 34.7\ °C$ and $T = 38.2\ °C$ without addition of surfactant ($\gamma = 0.00$). This implies, that the monomeric solubility limit of CO_2 in VP.PU 1431/glycerol/TCPP ($\psi_{1431} = 0.728$, $\psi_{glycerol} = 0.136$ and $\psi_{TCPP} = 0.136$) exceeds 10 wt% at $p = 200$ bar. Since less CO_2 has to be solubilized by the surfactant molecules this trend was somehow expected. However, the strength of the shift of the $\tilde{X}$ -point to lower surfactant mass fractions γ was striking and only understandable assuming a large monomeric solubility of CO_2 in the polyol mixture.

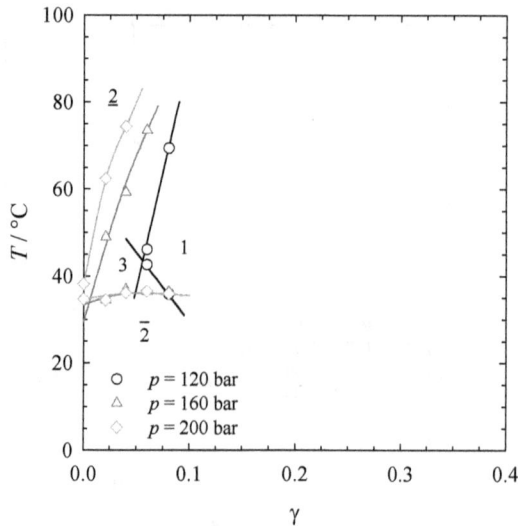

Figure 4.5: $T(\gamma)$-sections for the system VP.PU 1431/glycerol/TCPP $- CO_2 - $Q2-5211 ($\psi_{1431} = 0.728$, $\psi_{glycerol} = 0.136$, $\psi_{TCPP} = 0.136$) at $\alpha = 0.10$ and pressures of $p = 120$, 160 and 200 bar. With increasing pressure the mass fraction of the surfactant Q2-5211 needed to solubilize CO_2 in the polyol mixture decreases. At $p = 200$ bar the CO_2 is monomerically dissolved in the polyol mixture between $T = 34.7\ °C$ and $T = 38.2\ °C$.

The result that at $p = 200$ bar the entire CO_2 is monomerically dissolved, implies that the main idea of the POSME principle to provide a high number density of CO_2 nano-pools is not realized. Instead such a high monomeric solubility will lead to the nucleation of CO_2 bubbles in the polyol matrix during expansion. Thus, in order to create a high number density of CO_2 nano-pools the mass fraction of CO_2 was subsequently increased to $\alpha = 0.20$. The phase behavior of this system is shown in Figure 4.6. As can be seen, the optimum point $\tilde{X}$ is shifted to considerable larger surfactant mass fractions $\tilde{\gamma}$ and slightly higher temperatures upon increasing α. Again, the lower

phase boundary $\bar{2} \to 1$ is found to be independent on the pressure.

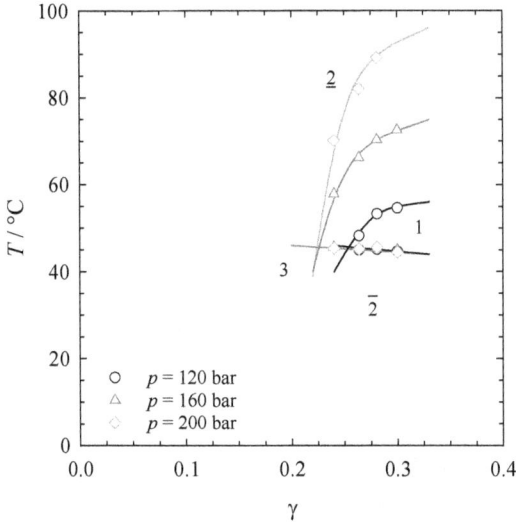

Figure 4.6: $T(\gamma)$-sections for the system VP.PU 1431/glycerol/TCPP – CO_2 – Q2-5211 ($\psi_{1431} = 0.728$, $\psi_{glycerol} = 0.136$, $\psi_{TCPP} = 0.136$) at $\alpha = 0.20$ and pressures of $p = 120$, 160 and 200 bar. With increasing the pressure from $p = 120$ bar to $p = 160$ bar the mass fraction of the surfactant Q2-5211 needed to solubilize CO_2 in the polyol mixture decreases.

In order to summarize the obtained results, the surfactant mass fraction $\tilde{\gamma}$ and temperature $\tilde{T}$ at the $\tilde{X}$-point are plotted as a function of the pressure (Figure 4.7). With increasing pressure the surfactant mass fraction $\tilde{\gamma}$ required to form a one-phase region monotonically decreases to a plateau value for each α-value. The plateau value decreases with decreasing mass fraction of CO_2 from $\tilde{\gamma} = 0.224$ at $\alpha = 0.20$ to $\tilde{\gamma} = 0.132$ at $\alpha = 0.15$. At $\alpha = 0.10$ the entire CO_2 is monomerically dissolved in the polyol mixture at $p \geq 200$ bar. Thus no surfactant is needed to form a one-phase mixture. Note that the trend observed with increasing pressure is similar to the trend found for balanced H_2O/NaCl – CO_2 – $F(CF_2)_iC_2H_4E_j$ microemulsions [41].

Considering the pressure-dependence of the phase inversion temperature $\tilde{T}$, at $\alpha = 0.15$ a monotonic increase is found with increasing pressure reaching a plateau at $\tilde{T} = 44.6$ °C. At $\alpha = 0.20$ the phase inversion temperature stays almost constant at $\tilde{T} \approx 45.7$ °C between

$p = 120$ bar and $p = 200$ bar. The almost constant value of $\tilde{T}$ can be explained considering the phase behavior of the binary systems. With increasing pressure, the upper miscibility gap of the $CO_2 - Q2$-5211 system shrinks, while the lower polyol $-$ Q2-5211 miscibility gap is unaffected and thus determines the location of the $\tilde{X}$ -point (compare Figure 4.2).

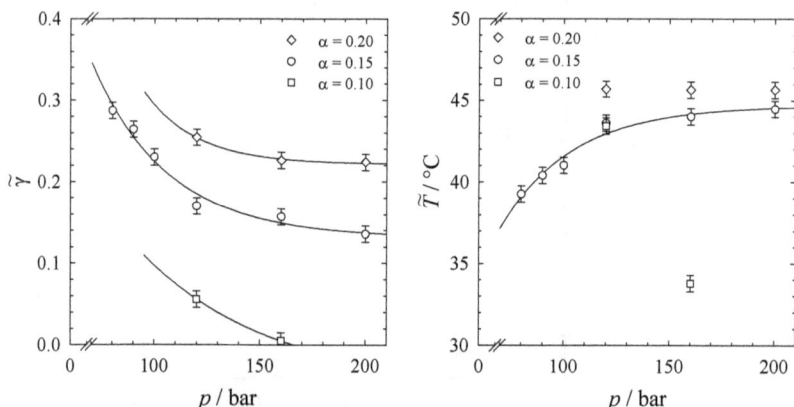

Figure 4.7: Surfactant mass fraction $\tilde{\gamma}$ (left) and temperature $\tilde{T}$ (right) at the optimum point $\tilde{X}$ of the microemulsion system VP.PU 1431/glycerol/TCPP $- CO_2 - Q2$-5211 ($\psi_{1431} = 0.728$, $\psi_{glycerol} = 0.136$, $\psi_{TCPP} = 0.136$) at $\alpha = 0.10$, 0.15 and 0.20 as a function of pressure. As the pressure increases, $\tilde{\gamma}$ monotonically decreases to a plateau for all α-values investigated. Simultaneously, at $\alpha = 0.15$ the temperature $\tilde{T}$ increase only slightly, while at $\alpha = 0.20$ a constant temperature value of $\tilde{T}$ was found. Note that the temperature $\tilde{T} = 33.8\,°C$ of the system at $\alpha = 0.10$ and $p = 160$ bar is determined by extrapolation, however, strongly deviates to lower T.

Having studied the phase behavior at different mass fractions α, the monomeric solubility of CO_2 in VP.PU 1431/glycerol/TCPP ($\psi_{1431} = 0.728$, $\psi_{glycerol} = 0.136$, $\psi_{TCPP} = 0.136$) can be determined from the $\tilde{\gamma}$ -values, if the phase inversion temperature stays almost constant. This prerequisite is fulfilled for $p = 120$ bar where $\tilde{T} = 44.6 \pm 1.2\,°C$. Thus, in Figure 4.8 the surfactant mass fraction $\tilde{\gamma}$ at the optimum point $\tilde{X}$ is shown as a function of α for $p = 120$ bar. The linear extrapolation of α to $\tilde{\gamma} = 0$ reveals a monomeric solubility of $\alpha_{mon} = 0.069$ ($p = 120$ bar) in VP.PU 1431/glycerol/TCPP. Note, that within the PuNaMi-project the monomeric solubility of CO_2 in VP.PU 1431 at $p = 120$ bar was determined by means of Raman experiments to $\alpha = 0.144$ at $T = 35\,°C$ and $\alpha = 0.126$ at $T = 60\,°C$ [222]. Thus the partial replacement of VP.PU 1431 by the polar glycerol and TCPP seem to lower the monomeric solubility of CO_2 in the polyol mixture considerably.

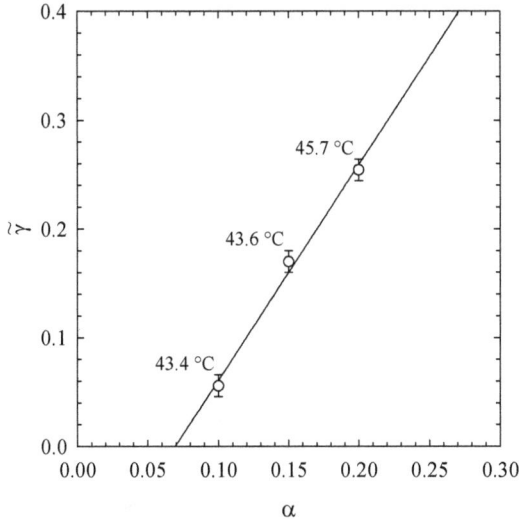

Figure 4.8: Surfactant mass fraction $\tilde{\gamma}$ at the optimum point $\tilde{X}$ of the microemulsion system VP.PU 1431/glycerol/TCPP – CO_2 – Q2-5211 ($\psi_{1431} = 0.728$, $\psi_{glycerol} = 0.136$, $\psi_{TCPP} = 0.136$) at $p = 120$ bar as a function of the CO_2 mass fraction α in the polyol/CO_2-mixture. The extrapolation of the α-value to $\tilde{\gamma} = 0$ allows to estimate the monomeric solubility of CO_2 in the polyol-mixture of 6.9 wt% at $p = 120$ bar. In addition the respective temperatures $\tilde{T}$ are specified.

4.1.3 Microstructure

In order to investigate whether the ternary system VP.PU 1431/glycerol/TCPP – CO_2 – Q2-5211 ($\alpha = 0.15$, $\psi_{1431} = 0.728$, $\psi_{glycerol} = 0.136$, $\psi_{TCPP} = 0.136$) is structured on the nano-scale and to determine the length scale, systematic small angle neutron scattering (SANS) measurements were performed.

As pointed out previously, the existence of CO_2 nano-pools in a polyol matrix is the main prerequisite of the POSME principle. Thereby also the size of these CO_2 nano-pools is of particular interest. Beside scattering methods (see section 2.3) the radius of the micelles swollen with CO_2 can be determined from the composition and geometrical considerations knowing the surface area a_C and volume v_C per surfactant molecule. Assuming spherical geometry, the radius R of monodisperse droplets is given by

$$R = 3\frac{v_C}{a_C}\frac{\phi_{C,i} + \phi_{CO_2,i}}{\phi_{C,i}}.$$

4.2

With $\phi_{C,i}$ being the volume fraction of surfactant in the internal interface and $\phi_{CO_2,i}$ being the volume fraction of CO_2 in the droplets. The volume per surfactant molecule v_C is defined as

$$v_C = \frac{M_C}{N_A \rho_C}.$$

4.3

Here $M_C \approx 721.1$ g mol^{-1} is the molar mass of the surfactant, N_A the Avogadro constant and $\rho_C = 1.02$ g cm^{-3} the macroscopic density of the surfactant Q2-5211. From surface tension measurements the surface area per surfactant molecule $a_C = 114.5$ Å^2 was determined (see Figure 4.3). The monomeric solubility of Q2-5211 in VP.PU 1431/glycerol/TCPP is equivalent to the cmc, which was determined to $\phi_{C,mon,a} = 0.022$. Thus, the volume fraction of surfactant in the internal interface is given by $\phi_{C,i} = \phi_C - \phi_{C,mon,a} \cdot (\phi_A + \phi_{CO_2,mon})$ with ϕ_A being the volume fraction of the polyol mixture. The volume fraction of CO_2 in the droplets $\phi_{CO_2,i}$ was calculated subtracting the volume fraction of CO_2 monomerically dissolved in VP.PU 1431/glycerol/TCPP $\phi_{CO_2,mon}$ from the entire volume fraction of CO_2. Using the surfactant mass fractions $\tilde{\gamma}$ at the optimum point $\tilde{X}$ ($p = 120$ bar, $\tilde{T} = 44.6 \pm 1.2$ °C) determined at $\alpha = 0.10$, 0.15 and 0.20 radii of $R = 5.8$, 4.6 and 4.5 nm are obtained, respectively (see Table 4.1). Note that the surface tension measurements and thus the determination of the monomeric solubility of Q2-5211 in VP.PU 1431/glycerol/TCPP was performed at $T = 21 \pm 1$ °C. The slightly higher radius at $\alpha = 0.10$ can be explained by either a higher monomeric solubility of CO_2 in the polyol mixture $\phi_{CO_2,mon}$ or a lower monomeric solubility of Q2-5211 in the polyol mixture $\phi_{C,mon,a}$.

Table 4.1: Volume fraction of CO_2 in the droplets $\phi_{CO_2,i}$, surfactant mass fraction $\tilde{\gamma}$ at the optimum point $\tilde{X}$, interfacial volume fraction of surfactant $\phi_{C,i}$ and radius R of monodisperse droplets in the ternary system VP.PU 1431/glycerol/TCPP $-$ CO_2 $-$ Q2-5211 ($\psi_{1431} = 0.728$, $\psi_{glycerol} = 0.136$, $\psi_{TCPP} = 0.136$) assuming spherical geometry. The radius was calculated from eq. 4.2 with $\phi_{C,mon,a} = 0.022$, $p = 120$ bar and $T = 44.6 \pm 1.2$ °C.

α	$\phi_{CO_2,i}$	$\tilde{\gamma}$	$\phi_{C,i}$	R / nm
0.10	0.039	0.056	0.043	5.8
0.15	0.081	0.170	0.167	4.6
0.20	0.113	0.254	0.250	4.5

In the next step, the microstructure and the length scale were studied by SANS-measurements. Thereby, not only the ternary system but also the corresponding binary system VP.PU 1431/glycerol/TCPP – Q2-5211 was studied in order to proof the formation of micelles in the polyol mixture. The original scattering data of the ternary ($\alpha = 0.15$, $\gamma = 0.22$, $p = 120$ bar) and binary ($\gamma_a = 0.25$, $p = 1$ bar) system are shown in Figure 4.9, left. As can be seen, the scattering intensity shows only a very little increase at low q-values and the incoherent scattering contribution dominates. This result was somehow expected, as the coherent scattering length densities of polyol, CO_2 and surfactant are very similar and thus the contrast very low. Subtracting the incoherent scattering contribution yields the coherent scattering intensity, shown in Figure 4.9, right. Thus, at high q-values a decrease of the scattering intensity proportional to q^{-4} is found. The slight shift of the scattering curve I_{meas}-I_{Incoh} towards lower q-values upon addition of CO_2 indicates a swelling of the structure by CO_2.

In addition a *Guinier* analysis ($I(q)$ versus q^2, Figure 4.9, right, inlet) was performed, which works without further parameter. A straight portion indicates the presence of spherical structures, with the slope being proportional to the radius of gyration R_g, i.e. $I(q) = I(0)\exp\left(-q^2 R_g^2/3\right)$ (see section 2.3.1). The radii obtained from the *Guinier* plots show that the surfactant micelles ($R_g = 1.4$ nm) swell by the addition of CO_2 ($R_g = 1.9$ nm). Due to the spreading of the data, which is caused by the small differences in the coherent scattering length density and the dominating incoherent scattering intensity, these values should be considered with some caution.

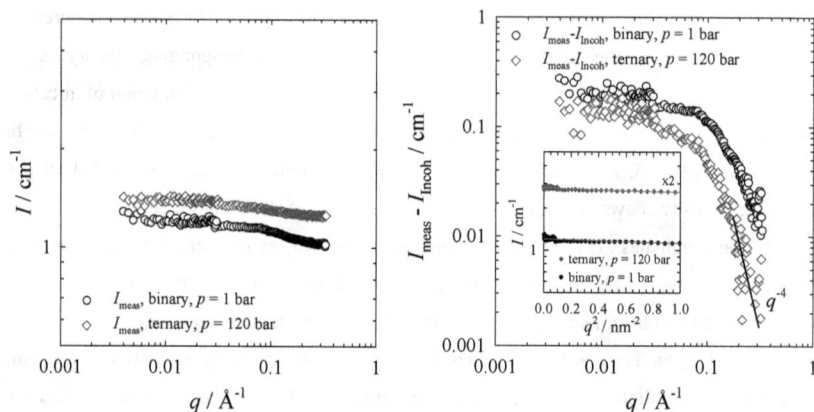

Figure 4.9: *Left*: Exemplary measured scattering intensity $I_{meas}(q)$ of the ternary system VP.PU 1431/glycerol/TCPP – CO_2 – Q2-5211 at $\alpha = 0.15$, $\gamma = 0.22$, $p = 120$ bar and $T = 44$ °C (blue diamonds) and the corresponding binary system VP.PU 1431/glycerol/TCPP – Q2-5211 at $p = 1$ bar and $T = 44$ °C (black circles). *Right*: Respective coherent scattering intensities obtained by subtracting the incoherent scattering contribution $I_{Incoh}(q)$. The *Guinier* plot for the binary (blue diamonds) and ternary system (black circles) is shown in the inlet. The slope of the straight portion indicates the presence of spherical structures and is proportional to the radius of gyration R_g [150, 163]. The position of the scattering curves and the radii obtained from the *Guinier* plots show that the surfactant micelles ($R_g = 1.4$ nm) swell by the addition of CO_2 ($R_g = 1.9$ nm). Note, that due to the spreading of the data, these values should be viewed with caution.

4.2 Proof of CO$_2$ nano-pools in polyol

In chapter 4.1 the phase behavior of the foamable CO$_2$-in-polyol system VP.PU 1431/glycerol/TCPP – CO$_2$ – Q2-5211 (ψ_{1431} = 0.728, $\psi_{glycerol}$ = 0.136, ψ_{TCPP} = 0.136) was studied in detail. Thereby a high monomeric solubility of CO$_2$ in the polyol mixture of 6.9 wt% was found at p = 120 bar and $\tilde{T}$ ≈ 44.6 ± 1.2 °C. Moreover, exploiting the data recorded by *Tseng* [218], it follows that the inverse phase sequence $\bar{2} \rightarrow 1 \rightarrow \underline{2}$ results from the phase behavior of the binary side systems. The phase boundaries and the location of the optimum point $\tilde{X}$ were found to be highly dependent on the pressure and therewith on the density of CO$_2$. First surface tension measurements indicate the formation of surfactant micelles in polyol. However, subsequent SANS-measurements are suffering from the low coherent scattering intensity due to the almost vanishing scattering length density contrasts.

In order to fully proof the existence of CO$_2$ nano-pools, the neutron scattering density contrast was enlarged using an at least partly deuterated polyol. The polyester polyol PES NFZ 6509, which exhibit an average molar mass of ~ 540 g mol^{-1} and a functionality of f = 3, turned out to be the polyol of choice. Using deuterated glycerol in the synthesis a partially deuterated polyol d-PES NFZ 6509 could be produced by Covestro, exhibiting a D/H ratio of ~ 0.262. In addition, 5.4 wt% of glycerol ($\psi_{glycerol}$ = 0.054) was added to the hydrophilic component to not only increase the polarity and thereby decrease the monomeric solubility of CO$_2$ in polyol, but also to allow to further increase the scattering contrast by using glycerol-d$_8$. Then, in the next step a one-phase microemulsion had to be formulated containing the new polyol/glycerol mixture.

4.2.1 Phase behavior

Systematic investigations showed that a one-phase microemulsion with PES NFZ 6509/glycerol – CO$_2$ can be formulated using a surfactant mixture of Q2-5211 and Genapol® T250p. Genapol® T250p is a non-ionic technical grade C$_{16/18}$E$_{25}$ surfactant. The respective $T(\gamma)$-sections of the system PES NFZ 6509/glycerol – CO$_2$ – Q2-5211/Genapol® T250p ($\psi_{NFZ\ 6509}$ = 0.946, $\psi_{glycerol}$ = 0.054, $\delta_{Genapol®}$ = 0.50) at α = 0.15 are shown in Figure 4.10 for 6 different pressure values between p = 80 and 200 bar. As can be seen, at intermediate surfactant mass fractions a one-phase region is found, while at small surfactant mass fractions a three-phase region appears. At high temperatures a polyol-rich microemulsion phase coexists with a CO$_2$-excess phase ($\underline{2}$). At low temperatures near the lower phase boundary an anisotropic two-phase coexistence is observed which might include the lamellar phase L_α. Note, that the lower phase boundary arises from the binary almost pressure-independent polyol – surfactant miscibility gap. Using the long-chain non-ionic $C_{16/18}E_{25}$ surfactant, which are known to form pronounced liquid crystalline phases in water, a generation of a lamellar phase also in polyol is likely [69]. Similar to the VP.PU 1431/glycerol/TCPP-microemulsion system the upper phase boundary $1 \rightarrow \underline{2}$ and consequently the optimum point $\tilde{X}$ shifts to lower surfactant mass fractions with increasing pressure. While at $p = 80$ bar a surfactant mass fraction of $\tilde{\gamma} = 0.169$ is required to form a one-phase region, at $p = 200$ bar only 11.3 wt% of the surfactant mixture is needed. Furthermore, the temperature of the optimum point $\tilde{X}$ shifts to slightly lower values.

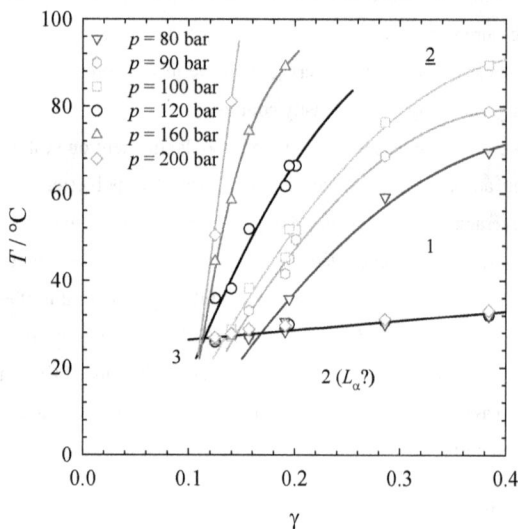

Figure 4.10: $T(\gamma)$-section through the phase prism of the microemulsion system PES NFZ 6509/glycerol – CO_2 – Q2-5211/Genapol® T250p ($\psi_{NFZ\,6509} = 0.946$, $\psi_{glycerol} = 0.054$, $\delta_{Genapol®} = 0.50$) at $\alpha = 0.15$ and 6 different pressures between $p = 80$ and 200 bar. As for the VP.PU 1431/glycerol/TCPP-microemulsion system, with increasing pressure the efficiency of the surfactant mixture Q2-5211/Genapol® T250p to solubilize CO_2 in the polyol mixture increases. At low temperatures near the pressure-independent lower phase boundary an anisotropic two-phase coexistence is observed which might include the lamellar phase L_α.

Thus, the surfactant mixture Q2-5211/Genapol® T250p ($\delta_{Genapol®} = 0.50$) seems to be slightly more efficient than the system VP.PU 1431/glycerol/TCPP – CO_2 – Q2-5211 ($\alpha = 0.15$, $\psi_{1431} = 0.728$, $\psi_{glycerol} = 0.136$, $\psi_{TCPP} = 0.136$). However, this efficiency increase might also be a consequence of a larger CO_2 monomeric solubility in the new polyol mixture.

The influence of the pressure on the optimum point $\tilde{X}$ ($\tilde{T}$, $\tilde{\gamma}$) is shown in more detail in Figure 4.11. With increasing pressure both the surfactant mass fraction $\tilde{\gamma}$ and temperature $\tilde{T}$ monotonically decrease to a plateau located at $\tilde{\gamma}$ = 0.109 and $\tilde{T}$ = 26.7 °C. In addition, $\tilde{\gamma}$ is plotted as a function of the density of CO_2 ρ_{CO_2} on a semi-logarithmic scale. The experimental errors amount to $\Delta p = \pm 5$ bar, the corresponding error in the CO_2 density $\Delta\rho_{CO_2}$ and $\Delta\tilde{\gamma}$ = ± 0.01. As found for the VP.PU 1431/glycerol/TCPP-microemulsion system, the lower surfactant mass fraction needed to solubilize CO_2 at higher pressure is a consequence of the higher CO_2-density and the higher monomeric solubility of CO_2 in the PES NFZ 6509/glycerol-mixture. Furthermore, the membrane theory suggests that the bicontinuous structured phase becomes instable, if the renormalized saddle splay modulus $\bar{\kappa}(\xi)$ becomes zero [92, 93]. Thus, the logarithm of the volume fraction of the surfactant residing in the amphiphilic film at the optimum point $\ln(\phi_{C,i,m})$ should be directly proportional to the bare saddle splay modulus $\bar{\kappa}_0$ [91–93]. As $\ln(\tilde{\gamma})$ turned out to be proportional to the CO_2-density, it can be concluded that $\bar{\kappa}_0$ should be also proportional to ρ_{CO_2}.

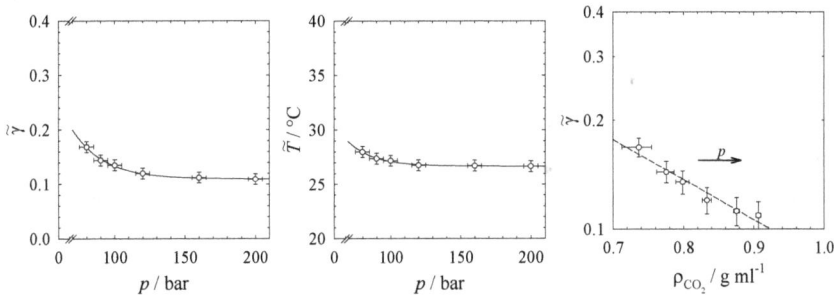

Figure 4.11: Surfactant mass fraction $\tilde{\gamma}$ (left) and temperature $\tilde{T}$ (middle) at the optimum point $\tilde{X}$ of the microemulsion system PES NFZ 6509/glycerol – CO_2 – Q2-5211/Genapol® T250p ($\psi_{NFZ\ 6509}$ = 0.946, $\psi_{glycerol}$ = 0.054, $\delta_{Genapol®}$ = 0.50) at α = 0.15 as a function of pressure. As the pressure increases, both $\tilde{\gamma}$ and $\tilde{T}$ decay to a plateau. Moreover, $\tilde{\gamma}$ is plotted as a function of the density of CO_2 in a semi-logarithmic scale (right). The experimental errors amount to $\Delta p = \pm 5$ bar, the corresponding error in the CO_2 density $\Delta\rho_{CO_2}$, $\Delta\tilde{\gamma}$ = ± 0.01 and $\Delta\tilde{T}$ = ± 0.5 K.

4.2.2 Microstructure

Replacing PES NFZ 6509/glycerol with the partially deuterated d-PES NFZ 6509/glycerol-d_8 the microstructure of the microemulsion system d-PES NFZ 6509/glycerol-d_8 – CO_2 –

Q2-5211/Genapol® T250p was systematically studied by means of SANS. Note, that the molar fractions were kept constant, i.e. $\psi_{\text{d-NFZ 6509}} = 0.942$, $\psi_{\text{glycerol-d}_8} = 0.058$. In more detail the microstructure of this system ($\alpha = 0.15$, $\delta_{\text{Genapol®}} = 0.50$) was studied as a function of composition ($\gamma = 0.18$ and 0.24), pressure ($p = 90$–200 bar) and temperature ($T = 24$–60 °C). Note, that due to the higher molecular mass of the deuterated components the composition parameters slightly change.

At first SANS-spectra of the binary system PES NFZ 6509/glycerol – Q2-5211/Genapol® T250p were recorded at $T = 40.6$ °C and $p = 1$ bar. Samples were prepared at 2 concentrations (dilute and concentrated) using the protonated PES NFZ 6509/glycerol mixture as well as the partly deuterated d-PES NFZ 6509/glycerol-d$_8$ mixture. The respective SANS-curves are shown in Figure 4.12, left ($\gamma_{\text{a,p}} = 0.059$ and $\gamma_{\text{a,d}} = 0.058$) and right ($\gamma_{\text{a,p}} = 0.205$ and $\gamma_{\text{a,d}} = 0.200$). In case of the diluted protonated system PES NFZ 6509/glycerol – Q2-5211/Genapol® T250p at $\gamma_{\text{a,p}} = 0.059$ (Figure 4.12, left, black diamonds) the incoherent scattering dominates. At low q-values only a slight increase of the scattering intensity is observed. Using the partly deuterated d-PES NFZ 6509/glycerol-d$_8$ mixture, the incoherent scattering contribution is reduced and the increase of the scattering intensity to low q-values much more pronounced. Note, that this increase at low q-values might be caused by the scattering from elongated structures or by critical fluctuations due to the possible proximity to the upper critical point of the miscibility gap (phase behavior of the pseudo-binary system not studied). These results suggest that the surfactant mass fraction at $\gamma_{\text{a,p}} = 0.059$ is only slightly above the *cmc*. In order to increase the scattering intensity the mass fraction of surfactant in the mixture of polyol and surfactant was increased to $\gamma_{\text{a,p}} = 0.205$ and $\gamma_{\text{a,d}} = 0.200$ (Figure 4.12, right). Thereby the scattering intensity at low to intermediate q-values increases strongly.

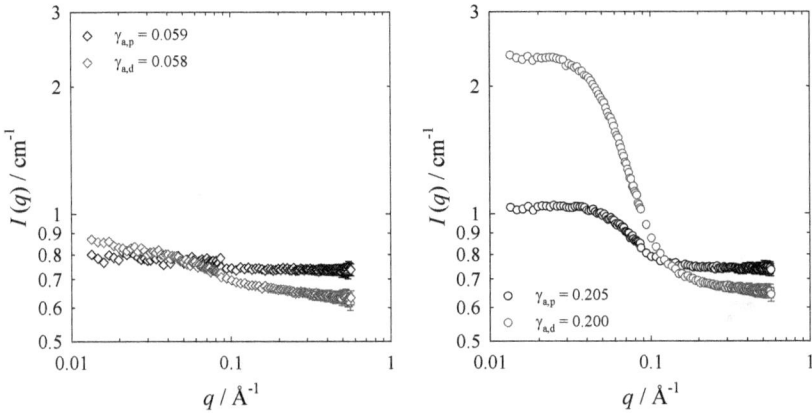

Figure 4.12: *Left*: Absolute scattering intensity I as a function of the scattering vector q of the protonated (PES NFZ 6509/glycerol, $\psi_{NFZ\ 6509} = 0.946$, $\psi_{glycerol} = 0.054$, black) and partly deuterated (d-PES NFZ 6509/glycerol-d_8, $\psi_{d\text{-}NFZ\ 6509} = 0.942$, $\psi_{glycerol\text{-}d_8} = 0.058$, blue) pseudo-binary system polyol – Q2-5211/Genapol® T250p ($T = 40.6\ °C$, $p = 1$ bar, $\delta_{Genapol®} = 0.50$), at low (left) and high surfactant mass fraction (right).

For a more quantitative analysis of the scattering curves the incoherent background is subtracted. The resulting scattering curves are shown in Figure 4.13. As can be seen, the SANS-curves show a constant scattering intensity at low q-values and a steep monotonic decay at intermediate to high q. Note, that a small dent at $q \approx \pi/R_0 \approx 0.14\ \text{Å}^{-1}$ can be observed only in the curve of the partly deuterated sample, indicating a large polydispersity. Note that a $\exp(-q^2 t^2)\cdot q^{-4}$-decay of the scattering intensity at high q-values indicates that the scattering length density profile is adequately described by a diffuse scattering length density profile.

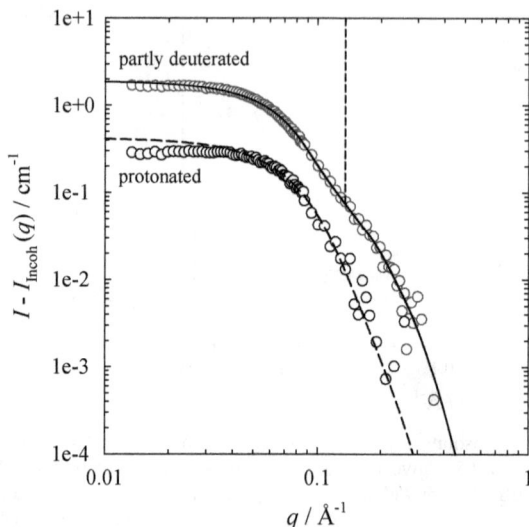

Figure 4.13: Coherent scattering intensity of the concentrated pseudo-binary protonated PES NFZ 6509/glycerol – Q2-5211/Genapol® T250p ($\gamma_{a,p}$ = 0.205, $\psi_{NFZ\,6509}$ = 0.946, $\psi_{glycerol}$ = 0.054) and partly deuterated d-PES NFZ 6509/glycerol-d_8 – Q2-5211/Genapol® T250p ($\gamma_{a,d}$ = 0.200, $\psi_{d\text{-}NFZ\,6509}$ = 0.942, $\psi_{glycerol\text{-}d_8}$ = 0.058) systems at $\delta_{Genapol®}$ = 0.50, p = 1 bar and T = 40.6 °C. The data were fitted using a combination of the form factor of polydisperse spheres [51, 203] and the *Percus-Yevick* structure factor [176, 177]. Note, that only core scattering was considered for the protonated system. The analysis of the protonated and partly deuterated system yields R_0 = 1.6 ± 1.0 nm, σ/R_0 = 0.63, R_{HS} = 2.1 nm and R_0 = 2.0 ± 1.1 nm, σ/R_0 = 0.55, R_{HS} = 3.2 nm, respectively.

The experimental data of both scattering curves were evaluated using a combination of the polydisperse spheres form factor and the *Percus-Yevick* structure factor. For the protonated system only core contributions were considered, while for the partly deuterated system contributions from the film, core and cross term were considered (see section 2.3.1). Consequently, the mean radius R_0 and hard sphere radius R_{HS} of the partly deuterated system, R_0 = 2.0 ± 1.1 nm and R_{HS} = 3.2 nm, respectively, are slightly larger than the radii of the protonated system R_0 = 1.6 ± 1.0 nm and R_{HS} = 2.1 nm. Note the large polydispersity indices of σ/R_0 = 0.55 and σ/R_0 = 0.63, respectively. The errors for the fit parameters were estimated to be less than 15%. The structural parameters are listed in Table 4.2. Note that the radius obtained from the parameter-free *Guinier* analysis via $R_g = \sqrt{3/5} \cdot R$, R_0 = 2.6 ± 0.1 nm, is slightly larger than the radii obtained from the fits.

Table 4.2: Structural parameters obtained from the analysis of the pseudo-binary protonated PES NFZ 6509/glycerol – Q2-5211/Genapol® T250p ($\gamma_a = 0.205$, $\psi_{NFZ\,6509} = 0.946$, $\psi_{glycerol} = 0.054$) and partly deuterated d-PES NFZ 6509/glycerol-d_8 – Q2-5211/Genapol® T250p ($\gamma_a = 0.203$, $\gamma_{a,d} = 0.200$, $\psi_{d\text{-}NFZ\,6509} = 0.942$, $\psi_{glycerol\text{-}d_8} = 0.058$) system at $\delta_{Genapol®} = 0.50$, $p = 1$ bar and $T = 40.6$ °C. The data were fitted using a combination of the form factor of polydisperse spheres [51, 203] and the *Percus-Yevick* structure factor [176, 177]. Note that only core scattering was considered for the protonated system. The thickness of the amphiphilic film and the ratio v_C/a_C were kept constant to $t = 5$ Å and $v_C/a_C = 12$ Å.

sample	$\phi_{C,i}$	$\Delta\rho_{core}$ / Å^{-2}	$\Delta\rho_{film}$ / Å^{-2}	R_0 / Å	σ / Å	σ/R_0	I_{Incoh} / cm^{-1}	ϕ_{disp}	R_{HS} / Å
prot.	0.180	$0.73 \cdot 10^{-6}$	-	16	10	0.63	0.747	0.080	21
partly deut.	0.180	$1.2 \cdot 10^{-6}$	$0.4 \cdot 10^{-6}$	20	11	0.55	0.672	0.180	32

In order to proof whether the added CO_2 dissolves monomerically in the polyol mixture or also swells the surfactant micelles the SANS-spectra of the ternary partly deuterated system d-PES NFZ 6509/glycerol-d8 – CO_2 – Q2-5211/Genapol® T250p ($\alpha = 0.15$, $\psi_{d\text{-}NFZ\,6509} = 0.942$, $\psi_{glycerol\text{-}d_8} = 0.058$, $\delta_{Genapol®} = 0.50$) were recorded at various pressures and temperatures at surfactant mass fractions of $\gamma = 0.18$ and 0.24. The coherent scattering intensity of this ternary system at $\alpha = 0.15$, $\gamma = 0.18$ and $p = 120$ bar is shown together with the corresponding pseudo-binary system d-PES NFZ 6509/glycerol-d8 – Q2-5211/Genapol® T250p ($\gamma_{a,d} = 0.200$, $p = 1$ bar) in Figure 4.14 at $T = 40.6$ °C.

Comparing the two scattering curves, it becomes obvious that the addition of CO_2 causes a shift of the curves to lower q-values proving the swelling of the micelles by CO_2.

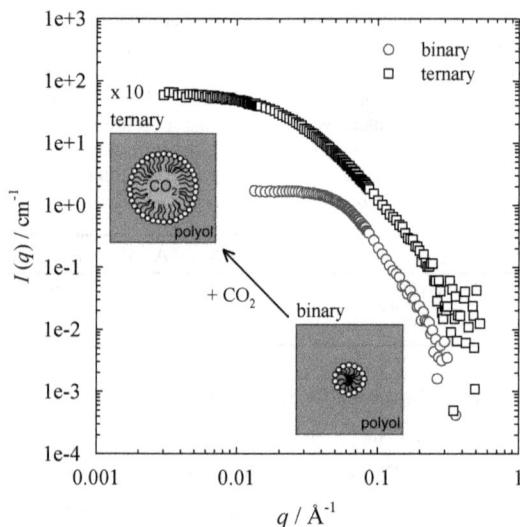

Figure 4.14: Coherent scattering curves of the partly deuterated binary system d-PES NFZ 6509/glycerol-d_8 – Q2-5211/Genapol® T250p ($\gamma_{a,d}$ = 0.200, $\psi_{d\text{-NFZ 6509}}$ = 0.942, $\psi_{glycerol\text{-}d_8}$ = 0.058, $\delta_{Genapol®}$ = 0.50, p = 1 bar, T = 40.6 °C) and the corresponding ternary system d-PES NFZ 6509/glycerol-d_8 – CO_2 – Q2-5211/Genapol® T250p (α = 0.15, γ = 0.18, T = 40.6 °C) at p = 120 bar. The addition of CO_2 causes a shift of the scattering curves to lower q-values, i.e. the structural length scale increases. The inlets show the swelling of the micelles by CO_2 schematically.

The influence of the temperature T on the SANS-curves is exemplary shown in Figure 4.15, left, for the two surfactant mass fractions γ = 0.18 and 0.24. The surfactant mass fractions γ and temperatures T, where SANS-curves were recorded, are marked by asterisks in the corresponding $T(\gamma)$-section shown in Figure 4.15, right. All SANS-curves show an almost constant or only slightly decreasing scattering intensity at low q-values. Then the intensity decreases monotonically to the incoherent background.

Considering at first the effect of temperature at γ = 0.24, the shoulder visible at intermediate q-values becomes less pronounced increasing the temperature from T = 40.6 °C to T = 60.0 °C. Furthermore, the SANS-curve shifts to lower q-values, indicating the formation of larger structures, which might be a consequence of the decreasing monomeric solubility of CO_2 in the polyol mixture. The sample at γ = 0.18 shows the reverse trend. At T = 60.1 °C a lower scattering intensity is observed at low q than at T = 40.7 °C. Looking at the phase diagram, one realizes that this scattering curve is recorded in the 2-phase region, where a microemulsion phase which contains less CO_2 coexists with a CO_2-excess phase.

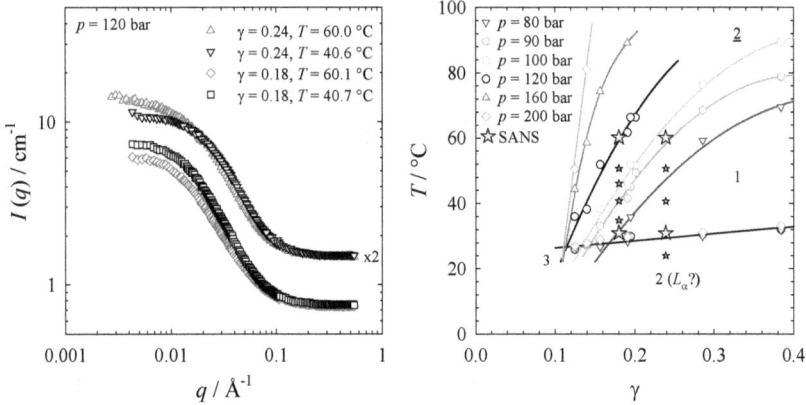

Figure 4.15: *Left:* Exemplary scattering curves of the system d-PES NFZ 6509/glycerol-d_8 – CO_2 – Q2-5211/Genapol® T250p   ($\alpha = 0.15$, $\psi_{\text{d-NFZ 6509}} = 0.942$, $\psi_{\text{glycerol-}d_8} = 0.058$, $\delta_{\text{Genapol®}} = 0.50$) at $p = 120$ bar. Note that at $\gamma = 0.18$, $T = 60.1$ °C the sample is studied in the 2-phase region, where a polyol-rich phase containing CO_2-swollen micelles coexists with a CO_2-excess phase. *Right:* Respective $T(\gamma)$-section for the protonated system PES NFZ 6509/glycerol – CO_2 – Q2-5211/Genapol® T250p ($\psi_{\text{NFZ 6509}} = 0.946$, $\psi_{\text{glycerol}} = 0.054$) at $\alpha = 0.15$ and 6 different pressures between $p = 80$ and 200 bar. The surfactant mass fractions γ and temperatures T of the SANS-experiments are marked by asterisks (the large symbols correspond to the scattering curves shown in Figure 4.15, left).

For a quantitative description of the scattering data a *Guinier* analysis was performed in the first step. In Figure 4.16, left, the *Guinier* and cross-section *Guinier* plot ($I(q) \cdot q^x$ vs q^2 for $x = 0, 1$ and 2) is exemplary shown for scattering data of the system at $\gamma = 0.18$, $p = 120$ bar and $T = 40.6$ °C. The straight line in the low q-regime when the exponent is set to $x = 0$ indicates the presence of spherical structures. Using the other exponents no linear behavior is found in the relevant q-regime (see section 2.3.1, eq. 2.48). Note, that for all partly deuterated polyol-rich CO_2-microemulsions studied in this thesis the *Guinier* plot is very similar. In Figure 4.16, on the right, the *Guinier* plots of the exemplary SANS-curves already discussed in Figure 4.15 are shown. The gyration radii obtained from the *Guinier* plots are in line with the qualitative discussion performed above. At $\gamma = 0.24$ the radius of gyration increases from $R_g = 5.2 \pm 0.1$ nm at $T = 40.6$ °C to $R_g = 8.0 \pm 0.3$ nm at $T = 60.0$ °C, while at $\gamma = 0.18$ the radius of gyration only increases from $R_g = 7.5 \pm 0.1$ nm at $T = 40.7$ °C to $R_g = 8.1 \pm 0.3$ nm at $T = 60.1$ °C.

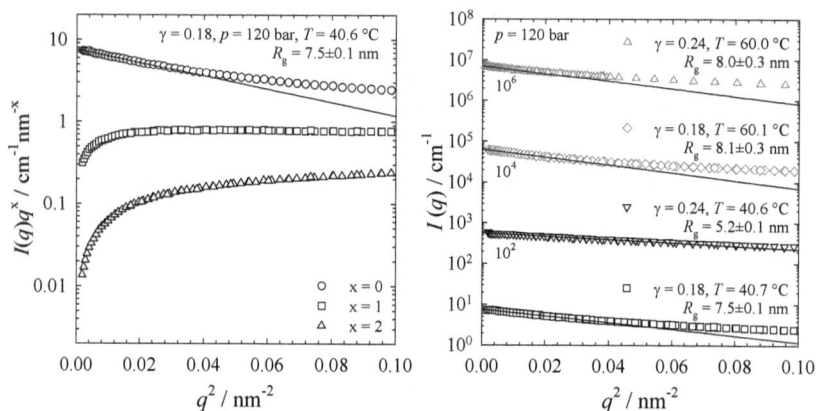

Figure 4.16: *Left:* Exemplary *Guinier* ($x = 0$) and cross-section *Guinier* ($x = 1$ or 2) plot [150, 163] for the ternary system d-PES NFZ 6509/glycerol-d_8 – CO_2 – Q2-5211/Genapol® T250p ($\alpha = 0.15$, $\psi_{\text{d-NFZ 6509}} = 0.942$, $\psi_{\text{glycerol-}d_8} = 0.058$, $\delta_{\text{Genapol®}} = 0.50$) at $\gamma = 0.18$, $p = 120$ bar and $T = 40.6$ °C. The slope of the straight portion for $x = 0$ indicates the presence of spherical structures and reveals a radius of gyration of $R_g = 7.5 \pm 0.1$ nm. *Right:* *Guinier* plots for the scattering data exemplary shown in Figure 4.15, left. Note, that the radius of gyration R_g increases with decreasing γ and increasing temperature.

The influence of the surfactant mass fraction, pressure and temperature on the radius of gyration (see Table 4.3) is shown in Figure 4.17. As can be seen, R_g decreases with increasing pressure (left) and decreasing temperature (right), respectively. This trend is most probably a consequence of two effects. Firstly, the density of CO_2 increases and thus the dispersed volume decreases with increasing pressure and decreasing temperature. Secondly, as shown in the PuNaMi-cooperation, the solubility of CO_2 in polyol increases with increasing pressure and decreasing temperature at the same time. Consequently, less CO_2 has to be dispersed by the surfactant with increasing pressure and decreasing temperature. Thus, both effects, related to the variation of the density and monomeric solubility of CO_2 in polyol act in the same direction, leading to a decrease of R_g with increasing pressure and decreasing temperature. Note, that the radius of gyration for the sample studied in the two-phase region 2 ($\gamma = 0.18$, $p = 120$ bar, $T = 60.1$ °C) is slightly smaller than predicted from the regression. This is due to the smaller CO_2 volume solubilized in the micelles, as a macroscopic CO_2-excess phase coexists with the CO_2-in-polyol microemulsion. Increasing the surfactant mass fraction the radius of gyration of the CO_2-swollen micelles decrease as the overall size of the amphiphilic film is enlarged.

Interestingly, by increasing the surfactant mass fraction the pressure dependency was found to slightly decrease, while the temperature dependency slightly increases.

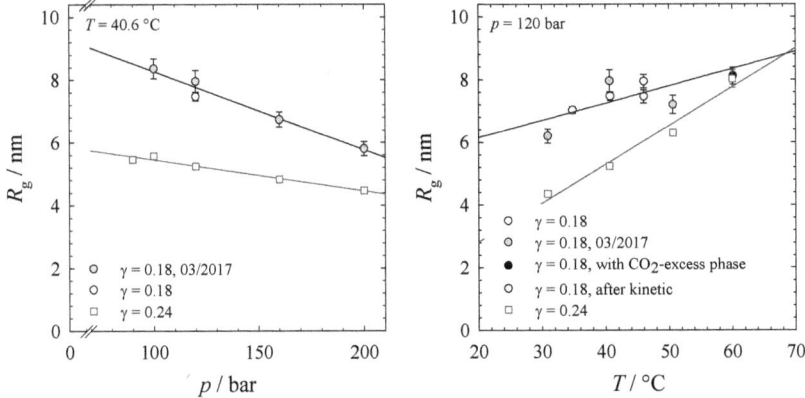

Figure 4.17: Radius of gyration R_g for the ternary system d-PES NFZ 6509/glycerol-d_8 –CO$_2$ – Q2-5211/Genapol® T250p ($\alpha = 0.15$) as a function of the pressure (*left*, at $T = 40.6$ °C) and temperature (*right*, at $p = 120$ bar) for various samples. Note that R_g decreases with increasing pressure, while increasing with temperature. The sample at $\gamma = 0.18$ was investigated during two beamtimes, revealing reproducible R_g-values. Moreover, at $\gamma = 0.18$, $p = 120$ bar and $T = 40.6$ °C (*right*) the SANS-spectra was recorded before (hollow circle) and after (yellow circle) the kinetic measurement, revealing a slight decrease of R_g, eventually due to a small CO$_2$-excess phase formed during the periodic pressure jumps into the 2-phase region. At $\gamma = 0.18$, $p = 120$ bar and $T = 60.1$ °C the system is in the two-phase region (2), while the microstructure of the lower polyol-rich microemulsion was studied.

The radii obtained from the *Guinier* analysis are summarized in Table 4.3.

Table 4.3: Radius of gyration R_g of the ternary system d-PES NFZ 6509/glycerol-d_8 – CO_2 – Q2-5211/Genapol® T250p ($\alpha = 0.15$, $\psi_{d\text{-NFZ 6509}} = 0.942$, $\psi_{glycerol\text{-}d_8} = 0.058$, $\delta_{Genapol®} = 0.50$) at various combinations of different surfactant mass fractions γ, temperatures T and pressures p obtained from the *Guinier* analysis [150, 163]. Note that the sample at $\gamma = 0.18$ was investigated during two beamtimes and using different batches of polyol. Furthermore in 04/2018 the kinetic piston was used, while the static piston was used for the samples measured in 03/2017 (marked with an *). As a consequence the mass fraction of CO_2, which is determined from the volume using the density and adjusting the position of the piston, might be slightly different (see Table 6.6). In addition, the samples at $\gamma = 0.24$, $T = 24.0$ °C and $\gamma = 0.18$, $T = 60.1$ °C were measured in the respective two-phase regions. The SANS-measurement [†] was performed after the pressure jump experiment.

γ	T / °C	p / bar	R_g / nm	ΔR_g / nm
0.18*	30.9	120	6.2	0.2
0.18*	40.6	100	8.4	0.3
0.18*	40.6	120	8.0	0.4
0.18*	40.6	160	6.7	0.2
0.18*	40.6	200	5.8	0.2
0.18*	50.6	120	7.2	0.3
0.18	34.8	120	7.0	0.1
0.18	40.7	120	7.5	0.1
0.18	46.0	120	7.9	0.2
0.18	60.1	120	8.1	0.3
0.18[†]	46.0	120	7.5	0.2
0.24	30.9	120	4.4	0.1
0.24	40.6	200	4.5	0.1
0.24	40.6	160	4.8	0.1
0.24	40.6	120	5.2	0.1
0.24	40.6	100	5.6	0.1
0.24	40.6	90	5.5	0.1
0.24	50.6	120	6.3	0.1
0.24	60.0	120	8.0	0.3

In order to gain insight into the type/shape of nanostructure formed in these foamable polyol-rich CO_2-microemulsions, additional molecular fragment dynamics (MFD)-simulations were performed within the PuNaMi-project by *H. Kuhn et al.* [223]. The simulations were performed near the optimal point $\tilde{X}$ of a similar polyol – CO_2 – Q2-5211 system at $\alpha = 0.15$,

$\gamma = 0.26$ and $p = 120$ bar. Since a different polyol was used the phase inversion temperature of this system is located at $T = 10\,°C$. The MFD-simulation results (Figure 4.18) indicate the formation of highly dynamic polyol- (blue) and CO_2-rich (green) domains with the trisiloxane surfactant (red/yellow) being located at the internal polyol/CO_2 interface. Locking at this snapshot, the structure resembles rather a connected bicontinuous structure than discrete CO_2-swollen micelles in a continuous polyol-matrix.

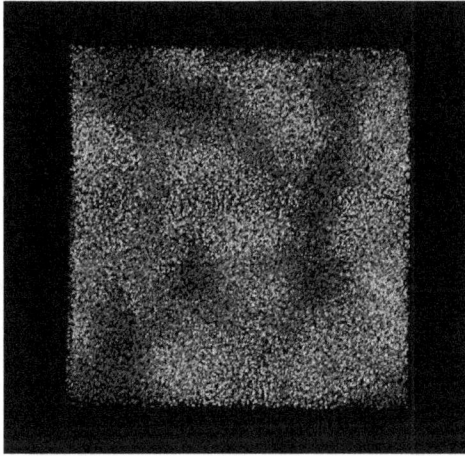

Figure 4.18: MFD-simulation near the optimal point $\tilde{X}$ of the system polyol $- CO_2 -$ Q2-5211 at $\alpha = 0.15$, $\gamma = 0.26$, $p = 120$ bar and $T = 10\,°C$ and box dimensions of $L_x = L_y = 50$ nm, $L_z = 100$ nm) performed by *H. Kuhn et al.* [223]. Blue: polyol, green: CO_2, red: polyoxyethylene chain of the surfactant, yellow: siloxane group of the surfactant. Note, that the simulation yields rather a connected bicontinuous structure than discrete CO_2-swollen micelles in a continuous polyol-matrix.

Based on these results the scattering curves were described by the *Debye* as well as *Teubner-Strey* model in the next step. While the *Debye* model describes the scattering of a system consisting of uncorrelated interfaces in a completely random system of a polar and a non-polar fraction via

$$I(q) = \frac{8\pi\langle\eta^2\rangle\xi_D^3}{\left(1 + \xi_D^2 q^2\right)^2} \qquad\qquad 4.4$$

with the characteristic length ξ_D [161], the *Teubner-Strey* model for bicontinuous structures

$$I(q) = \frac{8\pi c_2 \langle \eta^2 \rangle / \xi_{TS}}{a_2 + c_1 q^2 + c_2 q^4} .$$ 4.5

yields two length scales, i.e. the periodicity d_{TS} and correlation length ξ_{TS} from the combination of the parameter a_2, c_1 and c_2 (for details see section 2.3.1) [103]. In Figure 4.19, left, the scattering curve of the system d-PES NFZ 6509/glycerol-d_8 – CO_2 – Q2-5211/Genapol® T250p at $\gamma = 0.18$, $p = 120$ bar and $T = 40.6$ °C was described using both models. Note that the incoherent scattering background was subtracted from the scattering intensity.

As can be seen, both models describe the constant intensity at low q-values and the q^{-2}-decay at intermediate q-values. However, the scattering data at high q-values, are only quantitatively described by the *Teubner-Strey* model. As a consequence of the absence of the correlation peak, the parameter c_1 is positive and thus the periodicity d_{TS} is not defined and the amphiphilicity factor amounts to $f_a = 3$. However, due to the fact that the values for the amphiphilicity factor are $f_a > 1$, the disorder line ($f_a = 1$) is crossed, where the interfaces become uncorrelated and the quasiperiodical order is lost. Hence, the correlation function (eq. 2.34) becomes a simple exponential decay [151, 152]. Comparing the correlation lengths, $\xi_D = 2.8$ nm and $\xi_{TS} = 1.5$ nm, obtained from the analysis of the scattering curve, they turned out to differ by a factor of 2. However, using $\xi_{TS} = 2.8$ nm in the *Teubner-Strey* model both curves match. Interestingly, here the periodicity $d_{TS} = 1.9 \cdot 10^9$ nm, i.e. $d_{TS} \approx \infty$, and an amphiphilicity factor of $f_a = 1$ was obtained. The fact that $f_a = 1$ of the *Debye* model is smaller than the amphiphilicity factor obtained from the *Teubner-Strey* model with $\xi_{TS} = 1.5$ nm implies that the system is less structured.

The influence of pressure on the SANS-curves of the system at $\alpha = 0.15$, $\gamma = 0.18$, $T = 40.6$ °C and 4 pressures between $p = 100$ and 200 bar is shown in Figure 4.19, right. The parameters obtained from the evaluation of the scattering data with the *Teubner-Strey* model are listed in Table 4.4. The analysis yields correlation lengths of $\xi_{TS} = 1.44$ to 1.53 nm and an amphiphilicity factor f_a, which decreases with increasing pressure indicating an increasing ordering of the structure. Note that the correlation length ξ_{TS} increases with increasing pressure.

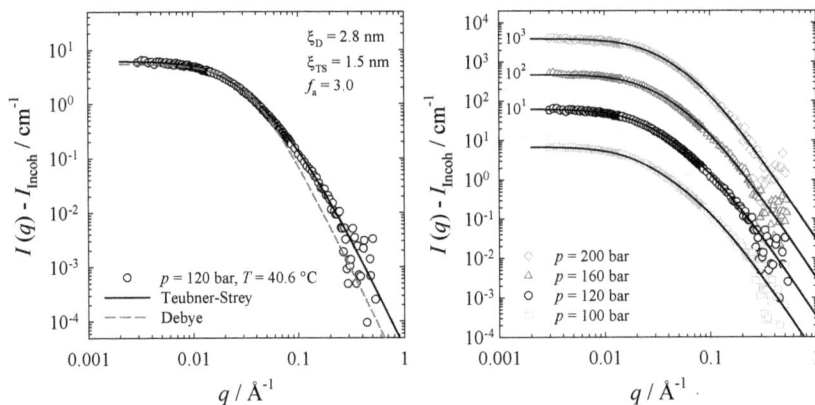

Figure 4.19: Coherent scattering curves of the system d-PES NFZ 6509/glycerol-d_8 – CO_2 – Q2-5211/Genapol® T250p ($\alpha = 0.15$, $\gamma = 0.18$). Please note that the scattering curve decrease monotonically, i.e. without a scattering peak. *Left*: The data recorded at $p = 120$ bar and $T = 40.6$ °C were evaluated using the *Teubner-Strey* [103] and *Debye* model [161]. Note that the *Debye* model ($\xi_D = 28$ Å, $8\pi<\eta^2> = 2.5 \cdot 10^{-4}$ Å^{-3}) does not quantitatively describe the data at low q-values. *Right*: Pressure dependence of the scattering curves recorded at $T = 40.6$ °C. The data were evaluated using the *Teubner-Strey* model for bicontinuous microemulsions.

Schubert and *Strey* described similar SANS spectra for near-critical bulk contrast samples, e.g. for the system D_2O/formamide-d_3 – n-octane – C_6E_2 ($\alpha = 0.369$, $\gamma = 0.47$, $\psi_{formamide} = 0.88$) [151]. Thus, they found similar values for the correlation length and amphiphilicity factor of $\xi_{TS} = 1.0$ nm and $f_a = 3.08$, respectively. However, please note the different oil mass fractions α.

Table 4.4: Fit parameters (correlation length ξ_{TS} and amphiphilicity factor f_a) obtained from the *Teubner-Strey* [103] model of the pressure and temperature dependent SANS-measurements of the system d-PES NFZ 6509/glycerol-d_8 – CO_2 – Q2-5211/Genapol® T250p ($\psi_{d-NFZ\ 6509} = 0.942$, $\psi_{glycerol-d_8} = 0.058$) at $\alpha = 0.15$ and $\gamma = 0.18$. Note that the periodicity d_{TS} is not defined due to the fact that $c_1 > 0$.

p / bar	T / °C	ξ_{TS} / nm	f_a
100	40.6	1.44	3.41
120	40.6	1.47	3.03
160	40.6	1.50	2.47
200	40.6	1.53	2.12

In order to elucidate the type and structure of phases coexisting at low temperatures another SANS-experiment was performed at $\gamma = 0.24$, $p = 120$ bar and $T = 24.0$ °C. Because the formed emulsion turned out to be rather stable, the SANS-measurement was performed without awaiting phase separation. Subtracting the incoherent background, the coherent scattering intensity is shown in Figure 4.20. As expected, the SANS-curve shows the typical strong decay of the scattering intensity at low q-values caused by the phase separation. Surprisingly, at intermediate q-values a pronounced minimum followed by a maximum at $q \approx \pi/R_0 \approx 0.05$ Å^{-1} is found, indicating the existence of fairly monodisperse structures. The $\exp(-q^2 t^2) \cdot q^{-4}$-decay of the scattering intensity at high q-values implies, that the scattering length density profile is adequately described by a diffuse scattering length density profile.

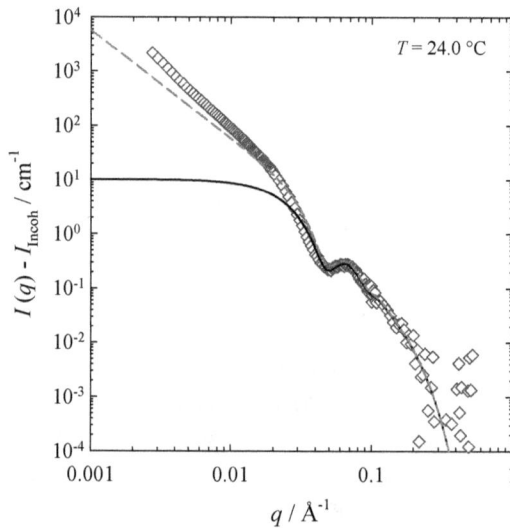

Figure 4.20: Coherent scattering intensity of the ternary system d-PES NFZ 6509/glycerol-d$_8$ – CO$_2$ – Q2-5211/Genapol® T250p at $\alpha = 0.15$, $\gamma = 0.24$, $p = 120$ bar and $T = 24.0$ °C obtained by subtracting the incoherent scattering contribution. Note that the SANS-measurement was performed in the two-phase region. The data were fitted using the form factor of the polydisperse spheres [170] with (red dashed line) and without (black solid line) the *Ornstein-Zernike* structure factor [166, 179]. Note that only film scattering was considered. The analysis of the scattering curve yields $R_0 = 5.9 \pm 1.2$ nm and $\sigma/R_0 = 0.20$.

The SANS-data were described using the form factor of polydisperse spheres (black solid line), considering only film contributions (see section 2.3.1). At intermediate and large q-values the scattering curve is quantitatively described, revealing a mean radius of $R_0 = 5.9 \pm 1.2$ nm with a small polydispersity index of $\sigma/R_0 = 0.20$. The structural parameters are listed in Table 4.5. Note that the mean radius is comparable to the gyration radii obtained from the *Guinier* plot (Table 4.3). The combination of the form factor with the *Ornstein-Zernike* structure factor, i.e. to model the scattering contribution caused by phase separation, almost describes the scattering data over the total q-range. However, the slope of the *Ornstein-Zernike* model is slightly smaller than the diminution of the scattering intensity.

Table 4.5: Structural parameters obtained from the analysis of the scattering curve of the system d-PES NFZ 6509/glycerol-d_8 – CO_2 – Q2-5211/Genapol® T250p ($\alpha = 0.15$, $\gamma = 0.24$, $\psi_{d\text{-NFZ }6509} = 0.942$, $\psi_{glycerol\text{-}d_8} = 0.058$, $\delta_{Genapol®} = 0.50$) at $p = 120$ bar and $T = 24.0$ °C. Note that the SANS-measurement was performed in the two-phase region. The data were fitted using a combination of the form factor of polydisperse spheres [170] and the *Ornstein-Zernike* structure factor [166, 179]. Note that only film scattering was considered.

v_C/a_C / Å	t / Å	$\phi_{C,i}$	$\Delta\rho_{film}$ / Å^{-2}	R_0 / Å	σ / Å	σ/R_0	I_{Incoh} / cm^{-1}	ϕ_{disp}	$S_{OZ}(0)$	ξ_{OZ} / Å
12	6	0.220	0.85	59	12	0.20	0.779	0.180	9.0·10^7	4.0·10^5

4.2.3 Expansion induced structural changes studied by TR-SANS

As has been shown above, we are now able to formulate foamable polyol-rich CO_2-microemulsions containing CO_2 nano-pools. Following the POSME procedure (Figure 1.2) the next step towards a nanocellular PU foam is the addition of isocyanate to the foamable microemulsion (to solidify the liquid foam via polymerization) which is followed by the expansion of the overall mixture. Because the synthesis of the PU foam would render the sapphire windows/ring cylinder of the pressure cells useless, the influence of sudden pressure changes on the microstructure is studied by means of the isocyanate-free polyol-rich CO_2-microemulsion combining periodic pressure jumps with time-resolved (TR-)SANS (described in section 6.3.2).

These kinetic studies were performed using the pseudo-ternary system d-PES NFZ 6509/glycerol-d_8 – CO_2 – Q2-5211/Genapol® T250p at $\alpha = 0.15$, $\gamma = 0.18$ and $T = 46.0$ °C. Starting at $p_{high} \approx 135$ bar, i.e. within the one-phase region, pressure jumps to a thermodynamically unstable state at $p_{low} = 53$ bar were performed, where a polyol-rich phase coexists with a CO_2-excess phase (2). The kinetic measurements were performed at a sample-to-detector distance of 20 m (the count rate at 39 m was too low to obtain an adequate signal-to-noise ratio in adequate time) adjusting $p_{high} = 135$ bar and $p_{low} = 53$ bar for $t_{high} = 9$ s and $t_{low} = 3$ s, respectively. Subdividing the pressure cycle into 119 slices, a time resolution of 0.1 s was achieved, while the remaining 0.1 s ensure the data transfer. Figure 4.21 shows the scattering curves as a function of the time in a 3D-plot (left) and the normalized integrated intensity nii, which is obtained from integration over the whole detector, as well as the pressure profile (right). The time, where the expansion was performed, was set to $t = 0$.

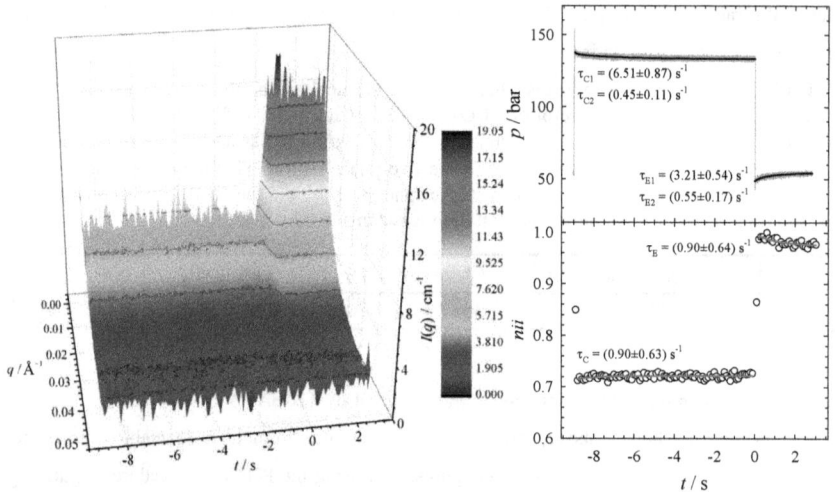

Figure 4.21: 3D-plot of scattering curves (*left*) and normalized integrated intensity (*nii*) as well as pressure (*right*) as a function of the cycle time t for the pseudo-ternary system d-PES NFZ 6509/glycerol-d_8 – CO_2 – Q2-5211/Genapol® T250p ($\alpha = 0.15$, $\gamma = 0.18$) at $T = 46.0$ °C. The pressure cycle, i.e. 9 s at $p_{high} = 135$ bar and 3 s at $p_{low} = 53$ bar, was subdivided into 119 slices with a time resolution of 0.1 s. The time at which the system was expanded from p_{high} to p_{low} was set to $t = 0$ (see pressure profile) and the hydraulic valves V1 and V2 were each opened for 0.05 s to adjust the respective pressure. The pressure cycle was repeated 250 times.

As can be seen in the 3D-plot a strong increase of the scattering intensity at low q-values is observed if the microemulsion is expanded to $p_{low} = 53$ bar, indicating the demixing of the sample. Moreover, the scattering length densities are dependent on the density of the respective components and thus contribute to the change of the scattering intensity in the pressure jump experiment. Mainly due to the strong increase of the scattering intensity at low q the normalized integrated intensity nii also increases from $nii \approx 0.72$ at p_{high} to $nii \approx 0.98$ at p_{low}. The fact, that only one nii-value between the compressed and expanded state could be recorded shows that the structural change of the microemulsion follows the pressure profile within the time resolution of 100 ms.

In a next step, the pressure profile was analyzed using a biexponential function. The slower process, described by the time constant $\tau_{C2/E2} = 0.45\text{--}0.55\ \text{s}^{-1}$ presumably corresponds to the heating or cooling time of the stroboscopic high-pressure (SHP) SANS cell after a quasi-adiabatic heating or cooling. In view of thermodynamics, the periodic experiments performed here are periodic quasi-adiabatic expansion and compression [46]. Whether an adiabatic cooling or heating is expected is difficult to predict, since the studied systems are complex mixtures structured at the nano-scale.[46] Furthermore, the *Joule Thomson* coefficients, μ_{JT}, of polyol (which is unknown) and CO_2 might have opposite signs under the adjusted experimental conditions ($T = 46.0\ °\text{C}$ and 53 bar $< p < 135$ bar), as H_2O and CO_2 exhibit values of $\mu_{JT}(H_2O) \approx -0.0208\ \text{K·bar}^{-1}$ and $\mu_{JT}(CO_2) = 0.901\text{-}0.156\ \text{K·bar}^{-1}$ [116]. Performing similar experiments on a water $-\ CO_2 -$ fluorinated surfactant microemulsion system *Müller et al.* have estimated a quasi-adiabatic heating of $\Delta T \approx +1$ K upon compression of $p = 180 \leftrightarrow 280$ bar at $T = 30.0\ °\text{C}$ [46]. The faster time constant is proposed to be a related to the adjustment of the respective solubility of CO_2 in the polyol mixture, which is found to depend strongly on the pressure. In the $nii(t)$ curves these processes cannot be resolved. Thus, the curves are analyzed by mono exponential function yielding an average time constant of $\tau_C = \tau_E = 0.90 \pm 0.64\ \text{s}^{-1}$.

An analogous kinetic measurement ($p_{high} = 135$ bar (9 s), $p_{low} = 53$ bar (3 s)) adjusting a time resolution of 0.05 s was performed. Thereby similar results were obtained, i.e. that the structural change of the microemulsion follows the pressure profile instantaneously, i.e. within a time resolution of 50 ms.

Table 4.6: Overview of the time constants obtained from the periodic pressure jump $p_{high} \leftrightarrow p_{low}$ ($t_{high} = 9$ s, $t_{low} = 3$ s) of the ternary system d-PES NFZ 6509/glycerol-d$_8$ $-\ CO_2$ $-$ Q2-5211/Genapol® T250p ($\alpha = 0.15$, $\gamma = 0.18$, $\psi_{NFZ\ 6509} = 0.946$, $\psi_{glycerol} = 0.054$, $\delta_{Genapol®} = 0.50$) at $T = 46.0\ °\text{C}$ to the thermodynamically unstable state.

p_{high} / bar	p_{low} / bar	$p(t)$		$nii(t)$	
		τ_C / s^{-1}	τ_E / s^{-1}	τ_C / s^{-1}	τ_E / s^{-1}
135	53	6.51 ± 0.87 0.45 ± 0.11	3.21 ± 0.54 0.55 ± 0.17	0.90 ± 0.63	0.90 ± 0.64

4.3 Influence of the co-oil on foamable polyol-rich CO_2-microemulsions

In this thesis foamable polyol-rich CO_2-microemulsions were formulated, in which CO_2-rich domains in a continuous polyol mixture are proven to exist using SANS (see section 4.2). First time-resolved SANS-measurements were performed during periodic pressure jumps in order to study the kinetics of phase separation, i.e. the change of the microstructure, upon expansion. These experiments show, that the onset of phase separation seems to follow the expansion instantaneously, while having adjusted a time resolution of $\Delta t = 50$ ms. The reason for the expansion induced phase separation are manifold. With decreasing pressure the solubility of CO_2 in polyol is decreased, thus more CO_2 has to be solubilized by the surfactant. Furthermore, the volume of the CO_2 increases due to the decreasing density. Both effects lead to a strongly increasing interface between polyol and CO_2. As the surfactant concentration stays constant the interfacial excess concentration decreases, which leads to a strong increase of the interfacial tension. As a consequence, aging phenomena as *Ostwald* ripening as well as coagulation followed by coalescence occur to minimize the interfacial energy. According to the Anti-Aging-Agent (AAA) concept suggested by *Strey et al.* [39] (see section 2.1.4), the use of a mixture of CO_2 and a low-molecular, hydrophobic co-oil was proposed to slow down the expansion-induced coarsening of the structure. Starting from the critical composition of CO_2/co-oil the mixture will separate spinodal, i.e. spontaneously, upon expansion. Due to the larger density of the co-oil and expected lower interfacial tension of co-oil and polyol, an enrichment of the co-oil at the interface is proposed. Consequently, such co-oils are expected to increase the stability of the CO_2-rich domains. In order to enable the incorporation in the PU matrix of the nanofoam co-oils with at least 2 OH-functionalities are favored. Thus, in this thesis 1,2-decanediol and 1,10-decanediol were chosen and their influence on the phase behavior, microstructure and phase separation kinetics of foamable CO_2-in-polyol microemulsion systems was studied. As a consequence of their molecular structure, 1,2-decanediol is expected to be more surface-active than 1,10-decanediol.

4.3.1 Phase behavior

The influence of the co-oils 1,2-decanediol and 1,10-decanediol on the phase behavior of ternary foamable CO_2-in-polyol systems was determined using the system HACH 175-3 – CO_2 – EP-R 357. HACH 175-3 is a polyol formulation consisting of different polyether and polyester polyols as well as glycerol. EP-R 357 is a technical trisiloxane surfactant ($M(D'E_j)M$) provided by Evonik Industries AG. It's composition is similar to Q2-5211. $T(\gamma)$-sections through the phase prism are shown in Figure 4.22 at $\alpha = 0.1275$ and pressures of $p = 80$ and 120 bar. As can be seen, the phase sequence $\overline{2} \rightarrow 1 \rightarrow \underline{2}$ which is typical for polyol-containing CO_2-microemulsions is observed also for this system. However, in contrast to the $T(\gamma)$-sections shown before, not only the upper but also the lower phase boundary is pressure dependent. Increasing the pressure from $p = 80$ bar to $p = 120$ bar results in a decrease of the surfactant mass fraction required for the formulation of a one-phase microemulsion from $\tilde{\gamma} = 0.137$ at $\tilde{T} = 29.5\ °C$ to $\tilde{\gamma} = 0.091$ at $\tilde{T} \approx 43.7\ °C$.

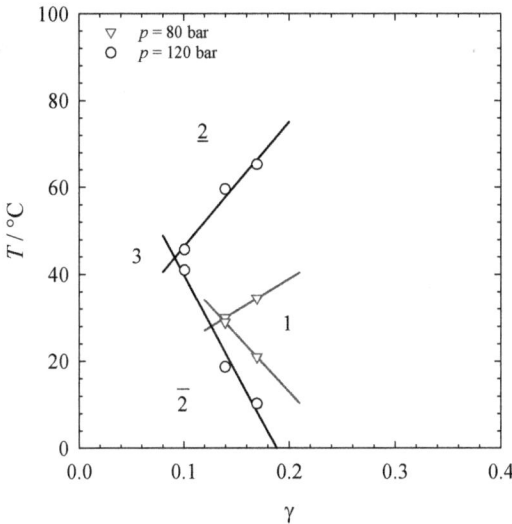

Figure 4.22: $T(\gamma)$-phase diagram of the system HACH 175-3 – CO_2 – EP-R 357 at $\alpha_{CO_2} = 0.1275$ and pressures of $p = 80$ and 120 bar. Again, with increasing pressure the efficiency of the surfactant EP-R 357 to solubilize CO_2 in the polyol increases. Please note that both the upper and the lower phase boundary are pressure-dependent.

In the next step, 15 wt% ($\beta = 0.15$) 1,2-decanediol and 1,10-decanediol, respectively, were added to the hydrophobic phase. The temperature dependent phase behavior of the system HACH 175-3 – CO_2/co-oil – EP-R 357 studied at 6 pressures between $p = 80$ and 200 bar is shown in Figure 4.23. Note, that the overall oil mass fraction α was set to $\alpha = (m_{CO_2} + m_{AAA})/(m_{CO_2} + m_{AAA} + m_{polyol}) = 0.15$, yielding a CO_2 mass fraction $\alpha_{CO_2} = (m_{CO_2})/(m_{CO_2} + m_{polyol}) = 0.1275$.

Looking at the phase diagrams, it becomes obvious, that the upper boundary $1 \rightarrow \underline{2}$ is independent of the co-oil used. Furthermore, compared to the system without co-oil, the upper phase boundary is only slightly shifted to higher temperatures. However, the lower phase boundary $\overline{2} \rightarrow 1$ is strongly shifted to lower temperatures adding 1,2-decanediol. Consequently, the efficiency of the surfactant EP-R 357 to solubilize CO_2 in HACH 175-3 increases significantly. When 1,2-decanediol is replaced by the less amphiphilic 1,10-decanediol the lower phase boundary $\overline{2} \rightarrow 1$, which results from the miscibility gap in the binary system polyol-surfactant, strongly shifts to higher temperature. As a result, the efficiency of the surfactant EP-R 357 to solubilize CO_2 in the polyol formulation decreases. These findings imply that both 1,2- and 1,10-decanediol partition between the interfacial film and polyol rather than acting as a co-oil. Thereby, particularly the amphiphilic 1,2-decanediol behaves as a co-surfactant.

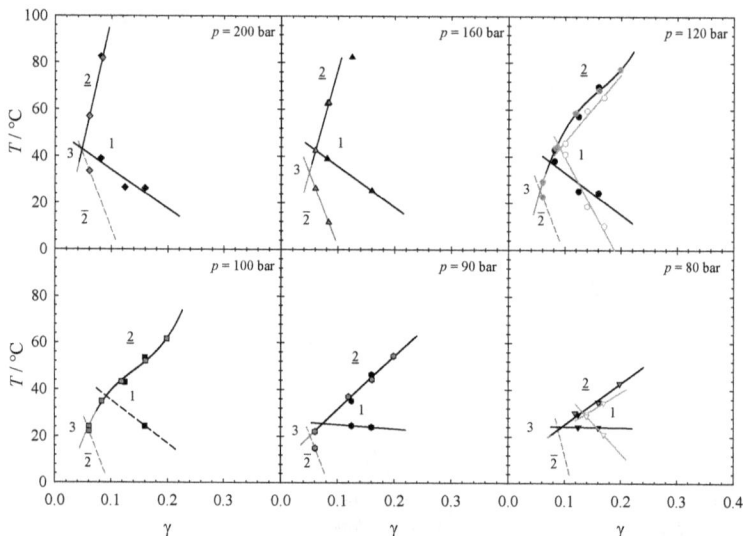

Figure 4.23: $T(\gamma)$-phase diagrams of the system HACH 175-3 – CO_2/co-oil – EP-R 357 at $\alpha_{CO_2} = 0.1275$, $\beta = 0.15$ and 6 pressures between $p = 80$ and 200 bar for 1,2-decanediol (red symbols), 1,10-decanediol (black symbols) and HACH 175-3 – CO_2 – EP-R 357 ($\alpha_{CO_2} = 0.1275$) at $p = 80$ and 120 bar (hollow symbols). Note that the upper phase boundary $1 \rightarrow \underline{2}$ is independent of the co-oil used. However, 1,2-decanediol strongly shifts the lower phase boundary to lower temperatures and therewith significantly increases the efficiency of the surfactant EP-R 357 to solubilize CO_2 in HACH 175-3, while 1,10-decanediol only slightly increases the efficiency.

4.3.2 Microstructure

In order to study the influence of co-oil on the microstructure of foamable nanostructured CO_2-in-polyol systems SANS-measurements were performed. Thus SANS spectra for the partly deuterated system d-PES NFZ 6509/glycerol-d_8 – CO_2/1,2-decanediol – Q2-5211/Genapol® T250p ($\psi_{\text{d-NFZ 6509}}$ = 0.942, $\psi_{\text{glycerol-}d_8}$ = 0.058, δ = 0.50) at α = 0.1725, α_{CO_2} = 0.15, β = 0.15, γ = 0.18 and T = 40.6 °C were studied. While detailed studies of the influence of the additive 1,2-decanediol on the phase behavior wasn't performed, the phase behavior of the SANS sample was checked prior to the beam time. In order to analyze the scattering curves shown in Figure 4.24, left, a *Guinier* analysis was performed, which revealed the presence of spherical structures. The pressure-dependent radii of gyration R_g obtained at T = 40.6 °C are plotted in Figure 4.24, right. For a better overview the R_g-values obtained from the 1,2-decanediol-free system at γ = 0.18 are also shown. The radii of the system d-PES NFZ 6509/glycerol-d_8 – CO_2/1,2-decanediol – Q2-5211/Genapol® T250p are summarized in Table 4.7.

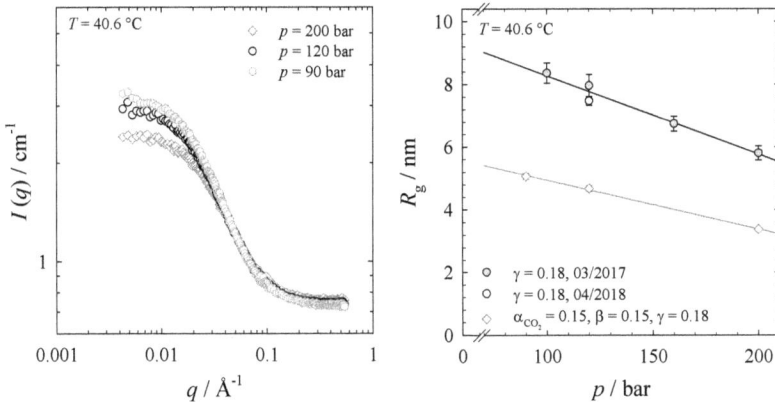

Figure 4.24: *Left*: Absolute scattering intensity I as a function of the scattering vector q of the partly deuterated system d-PES NFZ 6509/glycerol-d_8 – CO_2/1,2-decanediol – Q2-5211/Genapol® T250p (α = 0.1725, α_{CO_2} = 0.15, β = 0.15, γ = 0.18, $\psi_{\text{d-NFZ 6509}}$ = 0.942, $\psi_{\text{glycerol-}d_8}$ = 0.058, δ = 0.50) at T = 40.6 °C and 3 pressures of p = 90, 120 and 200 bar. *Right*: Radius of gyration R_g of the spherical structure for the partly deuterated system d-PES NFZ 6509/glycerol-d_8 – CO_2/1,2-decanediol – Q2-5211/Genapol® T250p ($\psi_{\text{d-NFZ 6509}}$ = 0.942, $\psi_{\text{glycerol-}d_8}$ = 0.058, δ = 0.50) as a function of the pressure at T = 40.6 °C. For a better overview, the gyration radii of the system without additive, i.e. d-PES NFZ 6509/glycerol-d_8 – CO_2 – Q2-5211/Genapol® T250p at α = 0.15 and γ = 0.18 are added.

As can be seen in general, the radius of gyration decreases with the addition of 1,2-decanediol. This finding supports the assumption that 1,2-decanediol acts both as co-solvent with respect to the polyol mixture and as a co-surfactant. While the former effect might lead to an increasing CO_2-solubility in polyol, the latter will increase the total interfacial area. Thus, both effects will lead to a decreasing radius. With respect to the limited beam time, at $p = 120$ bar SANS spectra were only recorded at $T = 40.6$ °C and $T = 46.0$ °C.

Table 4.7: Radius of gyration R_g of the ternary system d-PES NFZ 6509/glycerol-d_8 – CO_2/1,2-decanediol – Q2-5211/Genapol® T250p ($\alpha_{CO_2} = 0.1725$, $\alpha_{CO_2} = 0.15$, $\beta = 0.15$, $\gamma = 0.18$, $\psi_{d\text{-NFZ 6509}} = 0.942$, $\psi_{glycerol\text{-}d_8} = 0.058$, $\delta_{Genapol®} = 0.50$) at various combinations of different temperatures T and pressures p as obtained from the *Guinier* analysis [150, 163].

T / °C	p / bar	R_g / nm	ΔR_g / nm
40.6	200	3.4	0.1
40.6	120	4.7	0.1
40.6	90	5.1	0.1
46.0	120	4.4	0.1

4.3.3 Expansion induced structural changes studied by TR-SANS

In order to elucidate whether the additive 1,2-decanediol increases the stability of the CO_2-rich domains, microstructural changes due to sudden pressure changes were again studied combining periodic pressure jumps with TR-SANS. Thus, the polyol-rich CO_2/1,2-decanediol-microemulsion system d-PES NFZ 6509/glycerol-d_8 – CO_2/1,2-decanediol – Q2-5211/Genapol® T250p ($\alpha = 0.1725$, $\alpha_{CO_2} = 0.15$, $\beta = 0.15$, $\gamma = 0.18$, $T = 46.0$ °C, $\psi_{d\text{-NFZ 6509}} = 0.942$, $\psi_{glycerol\text{-}d_8} = 0.058$, $\delta = 0.50$) was exposed to periodic pressure jumps from $p_{high} = 132$ bar ($t_{high} = 9$ s), i.e. within the one-phase region, to the thermodynamically unstable state (2) at $p_{low} = 53$ bar ($t_{low} = 3$ s). Adjusting a time resolution of 0.1 s by subdividing the pressure cycle into 119 slices, 0.1 s were reserved to ensure the data transfer. Figure 4.25 shows the scattering curves as a function of the time in a 3D-plot (left) and the normalized integrated intensity nii as well as the pressure profile (right). The time, where the expansion was performed, was again set to $t = 0$.

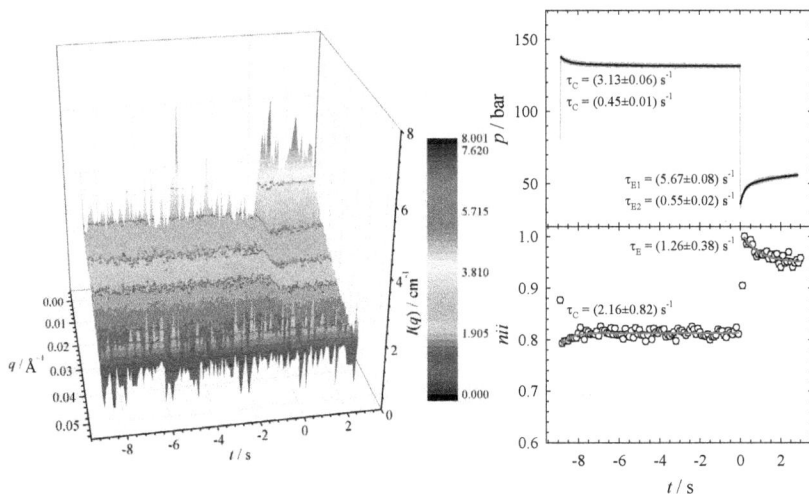

Figure 4.25: 3D-plot of scattering curves (*left*) and normalized integrated intensity *nii* as well as the pressure profile (*right*) as a function of the cycle time t of the pseudo-ternary system d-PES NFZ 6509/glycerol-d_8 – CO_2/1,2-decanediol – Q2-5211/Genapol® T250p ($\alpha_{CO_2} = 0.15$, $\beta = 0.15$, $\gamma = 0.18$) at $T = 46.0$ °C. The pressure cycle, i.e. 9 s at $p_{high} = 132$ bar and 3 s at $p_{low} = 53$ bar, was subdivided into 119 slices with a time resolution of 0.1 s. The time at which the system was expanded from p_{high} to p_{low} was set to $t = 0$ (see pressure profile). The pressure cycle was repeated 250 times.

As shown before, a strong increase of the scattering intensity at low q-values is observed if the microemulsion is expanded to $p_{low} = 53$ bar, indicating the demixing of the sample. However, compared to the system without co-oil not only the overall scattering intensity but in particular the scattering intensity at low q seems to be decreased. This might points to a decelerated coarsening of the microstructure. However, looking at the temporal development of the normalized integrated intensity (Figure 4.25, right) it can be seen, that *nii* increases instantaneously from $nii \approx 0.81$ at p_{high} to $nii = 0.95$ at p_{low}. As for the 1,2-decanediol-free system, only one *nii*-value between the compressed and expanded state could be recorded. Thus, the onset of the phase separation follows the pressure profile within the time resolution of 100 ms. However, the lower increase of the forward scattering intensity might indicate a slowing down of the coarsening process via the addition of 1,2-decanediol. Furthermore, the results of this work show that 1,2-decanediol partitions between the interfacial film and polyol rather than acting as a co-oil. Thus, 1,2-decanediol acts differently to hydrophobic co-oils, which should form a

CO_2/AAA-mixture in the droplets and should cause a spinodal separation of the phases during the expansion. Thereby the expected enrichment of the AAA at the interface in the expanded state is proposed to stabilize the foam following the AAA-concept suggested by *Strey et al.* [39].

As pointed out previously (section 4.2.3), one can distinguish between two processes in the pressure profiles characterized by the respective time constant, while the $nii(t)$ curves can only be analyzed by mono exponential functions yielding the time constant of one process (see Table 4.8).

Table 4.8: Overview of the time constants obtained from the periodic pressure jump $p_{high} \leftrightarrow p_{low}$ of the ternary system d-PES NFZ 6509/glycerol-d_8 – CO_2/1,2-decanediol – Q2-5211/Genapol® T250p ($\alpha = 0.1725$, $\alpha_{CO_2} = 0.15$, $\beta = 0.15$, $\gamma = 0.18$, $\psi_{NFZ\,6509} = 0.946$, $\psi_{glycerol} = 0.054$, $\delta_{Genapol®} = 0.50$) at $T = 46.0$ °C to the thermodynamically unstable state. For comparison, the constants for the 1,2-decanediol-free system ($\alpha = 0.15$, $\beta = 0.00$) are again listed. Note that for both systems holds that $t_{high} = 9$ s and $t_{low} = 3$ s.

β	α	p_{high} / bar	p_{low} / bar	$p(t)$		$nii(t)$	
				τ_C / s^{-1}	τ_E / s^{-1}	τ_C / s^{-1}	τ_E / s^{-1}
0.00	0.15	135	53	6.51 ± 0.87 0.45 ± 0.11	3.21 ± 0.54 0.55 ± 0.17	0.90 ± 0.63	0.90 ± 0.64
0.15	0.1725	132	53	3.13 ± 0.06 0.45 ± 0.01	5.67 ± 0.08 0.55 ± 0.02	2.16 ± 0.82	1.26 ± 0.38

5 Summary

The production of nanocellular materials as high performance thermal insulation materials is one major challenge in material science research and application. Promising candidates which already today exhibit a low thermal conductivity λ_{therm} are aerogels and vacuum insulation panels. However, both materials bear considerable weaknesses, with respect to cost, stability and continuous production. An alternative promising approach for the synthesis of polymeric nanocellular materials is the "Principle Of Supercritical Microemulsion Expansion" (POSME) proposed by *Strey, Sottmann* and *Schwan* in 2003 [27, 28]. During the last decade major efforts have been made to use the POSME procedure for the production of PU-nanofoams. One important prerequisite is the deep knowledge of the properties of foamable CO_2-microemulsions. Although some studies on the phase behavior, microstructure and kinetics of aqueous CO_2-microemulsions stabilized by a non-ionic fluorinated surfactant were performed [41, 45–51], and investigations elucidating the general features, e.g. the transition from CO_2-swollen micelles to bicontinuous structures and water-swollen micelles are so far missing.

In order to examine whether the general trends found for water – n-alkane – C_iE_j can be extended to CO_2-microemulsion systems, the microemulsion system $H_2O/NaClO_4 – CO_2 –$ Capstone® FS-3100 ($\varepsilon = 0.05$) was systematically studied. Thus, at first the phase behavior of the water-rich, balanced as well as CO_2-rich system was examined at pressures of $p = 150$–300 bar. Compared to the environmental harmful Zonyl® surfactants the phase behavior is located at significantly lower temperatures $\Delta\tilde{T} \approx 14$ K, while with increasing pressure the efficiency of the surfactant to solubilize water and CO_2 increases. However, the larger density of the liquid CO_2 causes a rather small pressure-dependence. Moreover, the Capstone® surfactant turned out to be slightly less efficient in solubilizing water and CO_2 than Zonyl® with respect to the density of CO_2 (Figure 3.21). In the balanced system at temperatures below the one-phase region a multiphase region containing a lamellar phase (L_α) was found (Figure 3.4).

Systematic small angle neutron scattering (SANS) experiments for the first time revealed a transition from CO_2-swollen cylinders in water (Figure 3.7) via a bicontinuous microemulsions (Figure 3.8) to water-swollen micelles in CO_2 (Figure 3.10). At constant surfactant mass fraction, the characteristic periodicity d_{TS} of the bicontinuous system increases with increasing pressure, due to the increasing monomeric solubility of the surfactant in CO_2. The scattering curve recorded within the "L_α present" region exhibits two scattering peaks described by amphiphilicity factors of $f_a = -0.786$ and $f_{a,L_\alpha} = -0.993$ (Figure 3.8). Being close to -1 the latter value proves the existence of a lamellar phase.

Both, the similar phase behavior and the temperature dependence of the characteristic length scale of aqueous CO_2- and classical water/oil-microemulsions suggest to scale the length scale onto one single curve. Such as scaling is indeed observed when the reduced characteristic length scale $\xi \cdot \phi_{C,i}$ is plotted as function of the reduced temperature $2(T-T_m)/T_u-T_l$ (Figure 3.11). Thus, the temperature dependence of the length scales of aqueous CO_2-microemulsions can be predicted when the surfactant volume fraction and temperature of the optimum point and the width T_u-T_l of the three-phase body are known.

Moreover, the bending rigidity of the amphiphilic film of the bicontinuous structure was studied by means of SANS and NSE. Even though translational diffusion as well as undulation modes are contributing on the same time scale, an almost quantitative agreement between the bending rigidity κ_{NSE} (via the simple *Zilman-Granek* approximation) and the renormalization corrected bending rigidity $\kappa_{0,SANS}$ (Figure 3.16) was revealed. In addition, the obtained bending rigidities agree reasonable well with results from other alkane as well as CO_2-microemulsions [41, 185, 211]. The numerical evaluation of the *Zilman-Granek* model, which is required for a more quantitative analysis, suffers from the fact that translational diffusion and fluctuation are on the same time scale (Figure 3.17).

Upon partial replacement of CO_2 by cyclohexane, an efficiency boosting was observed, i.e. the surfactant mass fraction needed to formulate an one-phase microemulsion is reduced by almost 50% (Figure 3.19 and Figure 3.20). Again a coexistence of lamellar and bicontinuous phase was found at low temperatures. Here, an anisotropic scattering pattern clearly proves the existence of the lamellar phase (Figure 3.22). Increasing the cyclohexane mass fraction the periodicity of the structure was found to slightly increase as the overall size of the amphiphilic film shrinks. The amphiphilicity factor f_a follows a parabola with a minimum i.e. highest ordering at about 8 wt% of cyclohexane in the CO_2/cyclohexane mixture. Consequently at this composition the largest bending rigidities were found (Figure 3.25). Unexpectedly, $\xi \cdot \phi_{C,i}$ was found to slightly increase linearly with increasing cyclohexane-content and decreasing pressure (Figure 3.24), which can be related to the pressure-dependent volume fraction of the CO_2/cyclohexane domains ϕ. However, from the performed NSE-measurements no reliable trend for the bending rigidity is observed, except for an increase of the bending rigidity κ_{NSE} with decreasing surfactant mass fraction.

In the second part of this thesis (section 4), the results gained for aqueous CO_2-microemulsions is used to formulate foamable nanostructured CO_2-in-polyol microemulsions for the production of nanocellular PU foams as high performance thermal insulation material. Starting point was the study of the binary system VP.PU 1431/glycerol/TCPP – trisiloxane surfactant (Q2-5211). Surface tension measurements (Figure 4.3) confirmed, that surfactant micelles in the polyol-rich component are formed, while due to the less polar and larger molecules the critical micelle concentration of the surfactant in the polyol-mixture is a factor of 250 larger than in H_2O. In order to study the influence of CO_2 and pressure on the phase behavior, temperature-dependent phase diagrams of the system VP.PU 1431/glycerol/TCPP $- CO_2 -$ Q2-5211 were recorded at different CO_2 mass fractions α and pressures between 80 and 200 bar (Figure 4.4, Figure 4.5 and Figure 4.6). With increasing pressure the surfactant mass fraction $\tilde{\gamma}$ required to form an one-phase microemulsion was found to monotonically decrease to a plateau. Comparing the values of $\tilde{\gamma}$ obtained for the different α-values, they are found to decrease strongly with decreasing mass fraction of CO_2, indicating a high monomeric solubility of CO_2 in polyol. The linear extrapolation of the $\tilde{\gamma}$-values to $\tilde{\gamma} = 0$ revealed a monomeric CO_2 solubility of 6.9 wt% at $p = 120$ bar and $\tilde{T} = 44.6 \pm 1.2$ °C (Figure 4.8). Having determined the monomeric solubility of CO_2 in the polyol-mixture and the critical micelle concentration the radius of the CO_2-swollen micelles under these conditions was determined from the composition and geometrical considerations to $R = 4.5–5.8$ nm (Table 4.1). The *Guinier*-analysis of first preliminary SANS-studies, which suffered from the very low scattering length density contrast and the dominating incoherent scattering contribution of the protonated system, indicated a little swelling of the structures from $R_g = 1.4$ nm to $R_g = 1.9$ nm when CO_2 is added (Figure 4.9).

In order to fully proof the existence of CO_2 nano-pools in a polyol matrix, the neutron scattering length density contrast was enlarged using a partly deuterated polyol. Note, that using a different polyol (which can be at least partly deuterated), a different surfactant mixture had to be developed to formulate a polyol-rich CO_2-microemulsion. Afterwards, the phase behavior of the system PES NFZ 6509/glycerol $- CO_2 -$ Q2-5211/Genapol® T250p ($\alpha = 0.15$) was studied in detail, revealing that with increasing pressure both the surfactant mass fraction $\tilde{\gamma}$ and temperature $\tilde{T}$ monotonically decrease to plateau-values of $\tilde{\gamma} = 0.109$ and $\tilde{T} = 26.7$ °C, respectively. Replacing PES NFZ 6509/glycerol with the partly deuterated d-PES NFZ 6509/glycerol-d_8 the microstructure of the pseudo-binary and pseudo-ternary system was systematically studied by means of SANS. Comparing the scattering curves and subsequent *Guinier*-analysis clearly proofs a swelling of the spherical structures by CO_2 from $R_g = 2.0 \pm 0.1$ nm to $R_g = 4.4-8.1$ nm (Figure 4.14). Studying the pressure and temperature dependence of the microstructure, it turned out, that the radius of gyration R_g (Figure 4.17) decreases with increasing pressure and decreasing temperature, in agreement with both the increasing CO_2 density and increasing CO_2 solubility in polyol. In addition, R_g is found to decrease when the overall size of the amphiphilic film is enlarged by increasing the surfactant mass fraction γ. Since MFD-simulations performed by *Kuhn* indicated the formation of highly dynamic continuous polyol- and CO_2-rich domains. The order of the SANS-spectra were analyzed with the *Teubner-Strey* structure factor in the next step revealing correlation lengths of the order of $\xi_{TS} = 1.5$ nm. Due to the absence of the correlation peak, the periodicity d_{TS} is not defined. Consequently, the amphiphilicity factor amounts to $f_a = 3$, i.e. indicating that the interfaces are uncorrelated and the quasiperiodical order is lost (Figure 4.19).

In order to simulate the expansion performed during the POSME principle, the influence of sudden pressure changes on the microstructure is studied by periodic pressure jumps combined with time resolved-SANS. It was found, that for pressure jumps from 135 bar (one-phase region) to 53 bar (2-phase region) the structural change of this polyol-rich CO_2-microemulsion follows the pressure profile instantaneously, i.e. faster than the time resolution of the experiment.

Following the AAA-concept suggested by *Strey et al.*, finally the influence of the additives 1,2-decanediol and 1,10-decanediol on the phase behavior of foamable nanostructured CO_2-in-polyol systems was studied. As predicted, the surfactant mass fraction $\tilde{\gamma}$ required to solubilize CO_2 in the polyol-mixture decreases upon addition of the additives (Figure 4.23). Moreover, the *Guinier*-analysis of the SANS-curves of the 1,2-decanediol-containing system yields smaller radii (Figure 4.24). These results suggest, that 1,2-decanediol partitions between the interfacial film and polyol rather than acting as a co-oil. In line with this, the structural changes induced by sudden pressure changes of the microemulsion are still faster than the time resolution of the time resolved-SANS experiment.

6 Appendix

6.1 Abbreviations and symbols

$\underline{2}$	index of an oil-in-water microemulsion with oil excess phase
$\overline{2}$	index of an water-in-oil microemulsion with water excess phase
$\tilde{\gamma}$	surfactant mass fraction at the $\tilde{X}$-point
$\tilde{T}$	temperature at the $\tilde{X}$-point
$\tilde{X}$	efficiency point at particular α
$\overline{\kappa}$	saddle splay modulus (also called Gauss modulus)
$\vec{k}_s$	wave vector of the scattered radiation
$\vec{k}_i$	wave vector of the incident radiation
$\langle \eta^2 \rangle$	average fluctuation of the scattering length density of two domains $\phi_a \phi_b \cdot \Delta \rho_s$
$(d\Sigma/d\Omega)_{H_2O}$	differential scattering cross section of water
$(d\Sigma/d\Omega)_{SA}$	differential scattering cross section of the sample
1	index of the one-phase state
3	index of the three-phase state
A	polar component (e.g. water, polyol)
a	prefactor for the characteristic length scale relation, ranging from $a = 4$–6
a	polar phase
a_2	parameter according to *Teubner-Strey*
a_{2,L_α}	parameter of the coexisting L_α-phase according to *Teubner-Strey*
AAA	Anti-Aging-Agent
a_C	head group area of a single surfactant molecule
A_d	surface area of a microemulsion droplet
$A_i(q)$	scattering amplitude of i
B	non-polar component (e.g. CO_2)
b	non-polar phase
B, B'	magnetic fields
B, B'	magnetic coils
b_i	coherent scattering length of the nuclei i
BMS	Bayer MaterialScience AG

BMWI	German Federal Ministry for Economic Affairs and Energy ("Bundesministerium für Wirtschaft und Energie")
C	amphiphilic component
c	prefactor for the description of the principal curvatures
c	concentration
c	surfactant-rich phase
c/w	CO_2-in-water
c_1	parameter according to *Teubner-Strey*
c_1, c_2	principal curvatures
c_{1,L_α}	parameter of the coexisting L_α-phase according to *Teubner-Strey*
c_2	parameter according to *Teubner-Strey*
c_{2,L_α}	parameter of the coexisting L_α-phase according to *Teubner-Strey*
cefb	CO_2-*emulsification failure boundary*
cep_α	critical end point α
cep_β	critical end point β
CFC	chlorofluorocarbon
C_iE_j	non-ionic n-alkyl polyglycolether
C_iG_j	non-ionic n-alkyl polyglycoside
cmc	critical micelle concentration
cp_α	critical point α
cp_β	critical point β
d	shell thickness
D	diffusion coefficient
d_g	diameter of the gas molecule
DLS	dynamic light scattering
d_s	characteristic length scale according to *Bragg*'s law
d_{SA}, d_{H_2O}	thickness of the sample and water, respectively
d_{TS}	characteristic periodicity according to *Teubner-Strey*
d_{TS,L_α}	characteristic periodicity of the coexisting L_α-phase according to *Teubner-Strey*
$d\sigma(q)/d\Omega$	differential scattering cross-section for homogeneous particles
efb	*emulsification failure boundary*
E_i	energy of the incident radiation

E_n	energy of the thermal neutrons
EO	ethylene oxide
E_S	surface energy
E_s	energy of the scattered neutrons
ESI-TOF-MS	electrospray ionization coupled with time-of-flight mass spectrometry
f	functionality
$F(CF_2)_iC_2H_4E_j$	polyfluoroalkyl polyglycolether
f_a	amphiphilicity factor
f_{a,L_α}	correlation length of the coexisting L_α-phase according to *Teubner-Strey*
F_b	free energy of the amphiphilic film
f_{corr}	correction factor for surface tension measurements
$f_{droplet}(r,R)$	radial density distribution function
F_{max}	maximum force during surface tension measurements
g_n	dimensionless magnetic moment (gyromagnetic ratio of the neutron)
H	mean curvature
h	Planck constant
H_0	spontaneous curvature
H12-MDI	1,1'-methylenebis(4-isocyanatocyclohexane)
HDI	hexamethylene diisocyanate
i	number of carbon-atoms in a C_iE_j or $F(CF_2)_iC_2H_4E_j$ surfactant
I_{BG}	environmental background intensity
I_{EC}	scattering intensity from the empty sample-cell
$I_{EC\_H_2O}$	scattering intensity from the empty cell of the calibration standard H_2O
I_{H_2O}	scattering intensity from the calibration standard H_2O
I_{Incoh}	incoherent scattering intensity
ILL	Institut Laue-Langevin
I_{SA}	scattering intensity from the sample
j	number of ethoxy units in a C_iE_j, $F(CF_2)_iC_2H_4E_j$ or $M(D'E_j)M$ surfactant or number of glycoside units in a C_iG_j surfactant
J_0	zeroth order *Bessel* function of first kind
K	*Gaussian* curvature
k	undulation wave vector
k_B	Boltzmann constant

k_{max}	upper cutoff undulation wave vector
k_{min}	lower cutoff undulation wave vector
L	cylinder length
l, l'	coil lengths
l_C	length of a single surfactant molecule
L_x, L_y, L_z	box dimensions of the MFD-simulations
L_α	lamellar phase
$M(D'E_j)M$	trisiloxane surfactant, with $M = Si(CH_3)_3O_{1/2}$-, $D' = -O_{1/2}Si(CH_3)(R)O_{1/2}$- and $R = -(CH_2)_3O$-
m_0, m_1, m_2, m_3	masses for the density measurement using a pycnometer
MDI	methylene diphenyl diisocyanate
$ME+L_\alpha$	index of a microemulsion phase coexisting with a lamellar phase
MFD	molecular fragment dynamics simulations
m_i	mass of the component i
M_i	molar mass of the component i
m_n	mass of a neutron
n	integer number
N_A	Avogadro constant
nanoPUR	titel of the cooperation between BMS/Covestro and University of Cologne/University Stuttgart
NBR 90	nitrile butadiene rubber of indentation hardness 90 (Shore)
ncb	near-critical boundary
NDI	naphthalene diisocyanate
NF-CID	Nanofoams by Continuity-Inversion of Dispersion
nii	normalized integrated intensity
NSE	Neutron Spin Echo
$\emptyset$	diameter
o/w	oil-in-water
oefb	*oil emulsification failure boundary*
p	pressure
$P(q)$	monodisperse particle form factor
$P(q,\tau(B))$	normalized intermediate scattering function
p_1, p_2	hydraulic pressures at pressure cycles
p_c	critical pressure (of CO_2)

$P_{cross}(q)$	cross sectional contribution to the form factor
$P_{cylinder}(q)$	polydisperse form factor for a cylinder
PFOA	perfluorooctanoic acid
p_g	gas pressure
p_{high}	upper pressure at pressure cycles (related to P1)
p_{inside}	pressure inside of a droplet
p_{low}	lower pressure at pressure cycles (related to P2)
PMDI	polymeric methylene diphenyl diisocyanate
PMMA	polymethylmethacrylat
PO	propylene oxide
POM	polyoxymethylene
POSME	Principle Of Supercritical Microemulsion Expansion
$p_{outside}$	pressure outside of a droplet
$P_{rod}(q)$	axial contribution to the form factor of a rod
PS	polystyrene
PU	polyurethane
PuNaMi	Polyurethane nano-cellular foams made from blowing agent based microemulsions for high performance thermal insulation
Q	invariant according to *Porod*
$q, \vec{q}$	absolute value of the scattering vector, scattering vector $q = |\vec{q}|$
R	gas constant
R_0	mean radius
R_1, R_2	principal curvature radii
R_c	core radius
R_g	radius of gyration
R_h	hydrodynamic radius
R_{HS}	hard sphere radius
RIM	reaction injection molding
R_r	ring radius for surface tension measurements
S	surface area of the amphiphilic film
$S(q)$	structure factor
$S(q,0)$	static structure factor
$S(q,\tau)$	dynamic structure factor

$S(q,\omega)$	scattering function
S/V	specific internal interface
SANS	Small Angle Neutron Scattering
SAXS	small angle X-ray scattering
sc	supercritical
SHP	stroboscopic high-pressure
$Si(x)$	sine integral function of x
$S_{OZ}(q)$	*Ornstein-Zernike* structure factor for critical fluctuations
$S_{PY}(q)$	*Percus-Yevick* structure factor
T	temperature
t	diffusivity
T_c	critical temperature (of CO_2)
TCPP	tris(2-chloroisopropyl) phosphate
t_D	time it takes a particle to diffuse a distance x_D
TDI	toluenediisocyanate
T_{EC}	transmission from the empty sample-cell
$T_{EC\_H_2O}$	transmission from the empty cell of the calibration standard H_2O
T_g	glass transition temperature
T_{H_2O}	transmission from the calibration standard H_2O
T_l	temperature of the critical end point α
T_m	mean temperature, i.e. $T_m = (T_u+T_l)/2$
TR	time resolved
T_{SA}	transmission from the sample
T_u	temperature of the critical end point β
T_α	temperature of the upper critical point α
T_β	temperature of the lower critical point β
V	volume
v	frequency of the pressure jumps
V1, V2	hydraulic ball valves for the respective hydraulic pressures p_1 and p_2
v_C	volume of a single surfactant molecule
V_i	volume of component i
VIP	vacuum insulation panel
V_m	molar volume

$V_{m,c}$	molar volume at the critical point (of CO_2)
v_n	velocity of a neutron
V_{part}	volume of the scattering particle
w/c	water-in-CO_2
w/o	water-in-oil
w_A	overall mass fraction of the component A (e.g. water)
$w_{A,max}$	maximum water mass fraction
$w_{A,p}$	overall mass fraction of the component A (e.g. water), related to the protonated system
w_B	overall mass fraction of the component B (e.g. CO_2)
$w_{B,p}$	overall mass fraction of the component B (e.g. CO_2) , related to the protonated system
wefb	*water emulsification failure boundary*
$W_{HS}(R_{HS}, R_{HS,0}, \sigma_{HS})$	*Gaussian* distribution function of the hard sphere radius R_{HS} around the mean value $R_{HS,0}$ with the distribution width σ_{HS}
$W_L(L)$	exponential distribution function of the cylinder length L with the average length $\langle L \rangle$
$W_R(R, R_0, \sigma)$	*Gaussian* distribution function of the radius R around the mean value R_0 with the distribution width σ
x	distance
$x(i)$	mole fraction of the component i
α	mass fraction of the non-polar phase in a mixture of polar and non-polar phase
α_{CO_2}	mass fraction of CO_2 in the mixture of polar and nonpolar components
α_n, α_n'	precession angle (see NSE)
α_p	mass fraction of the non-polar phase in a mixture of polar and non-polar phase, related to the protonated system
β	mass fraction of co-oil in a mixture of CO_2 and co-oil
β_i	effective mass fraction of co-oil in a binary mixture
β_{Kn}	constant of the *Knudsen* equation
β_{ZG}	*Zilman-Granek* exponent
β_σ	inverse activity coefficient
γ	overall surfactant mass fraction
$\gamma(r)$	correlation function
γ_0	mass fraction of surfactant monomers in the mixture of oil and water

γ_a	surfactant mass fraction in the polar phase
$\gamma_{a,p}$	surfactant mass fraction in the polar phase, related to the protonated system
γ_b	surfactant mass fraction in the non-polar phase
Γ_C	interfacial surfactant concentration
$\gamma_{C,mon,b}$	mass fraction of surfactant monomers in the hydrophobic component
γ_k	term for the *Zilman-Granek* relaxation rate
Γ_{max}	maximum surface excess
γ_p	overall surfactant mass fraction, related to the protonated system
Γ_{ZG}	*Zilman-Granek* relaxation rate
δ_i	mass fraction of the co-surfactant i in the surfactant/co-surfactant mixture
δ_p	characteristic size of pores
$\Delta p_{Laplace}$	*Laplace* pressure
ΔT	temperature extent of the three-phase body $\Delta T = T_u - T_l$
$\Delta \rho_{core}$	difference in the scattering length densities between the droplet core and the bulk phase, i.e. $\Delta \rho_{core} = \rho_{bulk} - \rho_{core}$
$\Delta \rho_{film}$	difference in the scattering length densities between the amphiphilic film and the bulk phase, i.e. $\Delta \rho_{film} = \rho_{bulk} - \rho_{film}$
$\Delta \rho_s$	scattering length density difference
ε	mass fraction of salt in the polar phase
ε_p	mass fraction of salt in the polar phase, related to the protonated system
ε_{ZG}	*Zilman-Granek* parameter, $\varepsilon_{ZG} \approx 1$
η_0	viscosity
θ	scattering angle
κ	bending rigidity modulus (also called mean bending modulus)
κ_0	renormalized corrected bending rigidity
$\kappa_{0,SANS}$	renormalized corrected bending rigidity as obtained from SANS
κ_{NSE}	bending rigidity as obtained from NSE, equals $\kappa_{0,SANS}$
λ	wavelength
λ_g	gaseous thermal conductivity
$\lambda_{g,0}$	free air conduction of a gas
λ_i	wavelength of the incident radiation
λ_{IR}	heat transfer via infrared radiation
λ_m	thermal conductivity along the solid matrix
λ_s	wavelength of the scattered radiation

λ_{therm}	thermal conductivity
μ	nuclear magneton
μ_{JT}	*Joule-Thomson* coefficient
ξ	characteristic length scale of the microemulsion structure
ξ_D	correlation length according to the *Debye* model
ξ_m	maximal length scale of the microemulsion structure, i.e. characteristic length scale at the $\tilde{X}$-point
ξ_{OZ}	*Ornstein-Zernike* correlation length of critical scattering
ξ_{TS}	correlation length according to *Teubner-Strey*
ξ_{TS,L_α}	amphiphilicity factor of the coexisting L_α-phase
ρ_{bulk}	scattering length density of the continuous medium surrounding a structure
ρ_c	critical density (of CO_2)
ρ_{core}	scattering length density of a structures core
ρ_{film}	scattering length density of a structures amphiphilic film
ρ_i	macroscopic density of the component i
$\rho_{s,domain}$	scattering length density of a domain
$\rho_{s,k}$	scattering length density of component k
σ	polydispersity of R_0
σ_0	surface tension of the pure solvent
σ_a	surface tension of the hydrophilic component with an amphiphile
σ_{ab}	interfacial tension between the hydrophilic and hydrophobic component
σ_{cmc}	surface tension at the *cmc*
σ_{HS}	polydispersity of a hard sphere
τ, $\tau(B)$	*Fourier* time (as function of the magnetic field)
τ_C, τ_{C1}, τ_{C2}	time constant for the process in the compressed state (p_{high})
τ_E, τ_{E1}, τ_{E2}	time constant for the process in the expanded state (p_{low})
ϕ	volume fraction of oil in the oil/water-mixture
$\phi_{C,i}$	volume fraction of surfactant at the internal interface
$\phi_{C,i,m}$	volume fraction of surfactant at the internal interface at the $\tilde{X}$-point
$\phi_{C,mon}$	volume fraction of surfactant monomerically solubilized
$\phi_{C,mon,a}$	volume fraction of surfactant monomerically solubilized in a
$\phi_{C,mon,b}$	volume fraction of surfactant monomerically solubilized in b
$\phi_{CO_2,i}$	internal volume fraction of CO_2, monomeric solubility considered

$\phi_{CO_2,mon}$	volume fraction of CO_2 monomerically solubilized
ϕ_{disp}	volume fraction of the dispersed phase
ϕ_i	volume fraction of the component i
χ	diffusivity parameter of the amphiphilic film
χ_T	isothermal compressibility
ψ_i	mass fraction of additive i to the hydrophilic phase
$\omega(k)$	dispersion relation of fluctuating membranes in a viscoelastic fluid
$\hbar$	reduced Planck constant ($\hbar = h/2\pi$)
$\hbar\omega$	energy transfer during inelastic scattering experiments

6.2 Chemicals

The substances used in this work are listed in Table 6.1 with the chemical formula, source of supply, molar mass and purity if available.

Capstone® FS-3100 was purified prior to use. The surfactant was solved in ethanol, dried over Na_2SO_4, filtrated and the ethanol was removed with a vacuum rotary evaporator.

The EO-polyester polyols Desmophen® VP.PU 1431 ($f = 2$, $M \approx 370\ \mathrm{g\ mol^{-1}}$) and PES NFZ 6509 ($f \approx 3$, $M \approx 540\ \mathrm{g\ mol^{-1}}$) as well as the polyol mixture HACH 175-3 were provided by Covestro. Thereby, in general, the technical synthesis yields highly technical grade polyols, i.e. a polydisperse mixture of different polyols with a broad molar mass distribution. Being more precise, the synthesis of the polyester polyol Desmophen® VP.PU 1431, used in this work, starts from a dicarboxylic acid and ethylene glycol. The molar mass distribution of Desmophen® VP.PU 1431 was studied by means of electrospray ionization (positive ion polarity) coupled with time-of-flight mass spectrometry (ESI-TOF-MS), which yields the intensity as a function of the molecular mass divided by the charge m/z, shown in Figure 6.1. As can be seen, the spectrum exhibits various peaks at m/z-values of 133 to 1042 u showing the polydispersity of the polyol. Note that several peaks have a deviation of $\Delta m/z = 44$ u from each other, which corresponds to the molar mass of one ethylene oxide unit (C_2H_4O, $M = 44.05\ \mathrm{g\ mol^{-1}}$).

HACH 175-3 is a polyol mixture of different polyether and polyester polyols as well as glycerol.

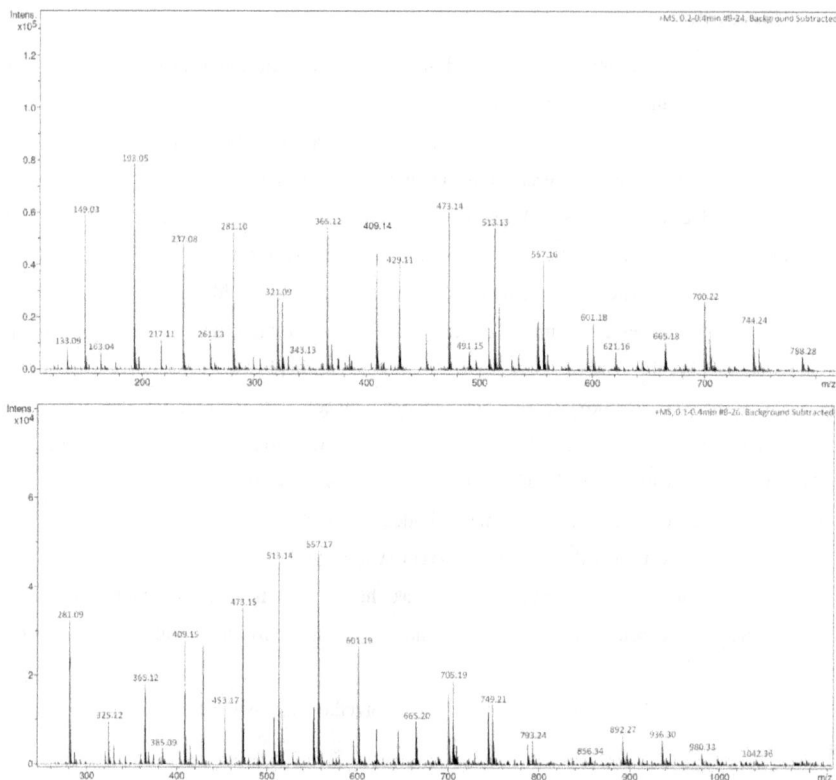

Figure 6.1: Intensity as function of the molecular mass divided by the charge m/z of Desmophen® VP.PU 1431 determined by means of electrospray ionization (positive ion polarity) coupled with time-of-flight mass spectrometry (ESI-TOF-MS). Scan regions of $\Delta m/z = 50$–1000 u (top) and $\Delta m/z = 250$–3000 u (bottom) were adjusted. Note that the broad molar mass distribution ($m/z \approx 133$–1042 u) is caused by the technical synthesis. Moreover, a characteristic difference of $\Delta m/z = 44$ u is found often, which is related to the molar mass of one ethylene oxide unit (C_2H_4O, $M = 44.05$ g mol^{-1}).

Table 6.1: List of substances used in this work.

substance	chemical formula	source of supply	molar mass / $g \cdot mol^{-1}$	purity / %
1,10-decanediol	$C_{10}H_{22}O_2$	Tokyo Chemical Industry	174.28	> 95.0
1,2-decanediol	$C_{10}H_{22}O_2$	Tokyo Chemical Industry	174.28	> 99.0
$C_{10}E_4$	$C_{18}H_{38}O_5$	BioChemika	334.49	> 97.0
Capstone® FS-3100	$F(CF_2)_iC_2H_4E_j$	DuPont Lehmann&Voss&Co.	-	tech. pure
carbon dioxide	CO_2	Linde / Westfalen	44.01	> 99.5
cyclohexane	C_6H_{12}	Merck	84.16	> 99.5
Desmophen® VP.PU 1431	-	Covestro Deutschland AG	~ 370	tech. pure
deuterium oxide	D_2O	Eurisotop	20.03	99.9
d-PES NFZ 6509	-	Covestro Deutschland AG	~ 580	tech. pure
EP-R 357	$M(D'E_j)M$	Evonik AG	-	tech. pure
Genapol® T250p	$C_{16/18}E_{25}$	Clariant	-	tech. pure
glycerol	$C_3H_8O_3$	Brenntag GmbH	92.09	99.8
glycerol-d$_8$	$C_3D_8O_3$	Sigma Aldrich	100.14	98.0
HACH 175-3	-	Covestro Deutschland AG	-	tech. pure
PES NFZ 6509	-	Covestro Deutschland AG	~ 540	tech. pure
Q2-5211 Xiameter™ OFX 5211	$M(D'E_{10})M$	Dow Corning	~ 721.1	tech. pure
sodium chloride	$NaCl$	Fluka	58.44	> 99.5
sodium perchlorate	$NaClO_4$	Sigma Aldrich	120.44	98.0
TCPP	$C_9H_{18}Cl_3O_4P$	BCD Chemie GmbHNL	327.55	tech. pure
water	H_2O	-	18.02	2x dist.

6.3 Experimental methods

6.3.1 Phase behavior measurements

The phase behavior measurements were carried out using a home-built high-pressure view cell designed by *Schwan* [107] and the mechanical workshop of the Institute of Physical Chemistry, Cologne. A 3D-rendered cross section of the cell is shown in Figure 6.2.

Figure 6.2: 3D-rendered cross section of the home-built high-pressure view cell, which was used to study the phase behavior. The sapphire ring cylinder (highlighted in blue) is sealed via two sealing rings by a screw driven piston and a block made of brass at the bottom. A small capillary connects the sample volume to a needle valve, which is used for filling and cleaning of the cell. The pressure is adjusted by the piston and controlled by a miniature pressure transducer. [107] – Reproduced by permission of The Royal Society of Chemistry

A polished sapphire ring cylinder (50 mm in height, $\varnothing_{outside}$ = 32 mm, $\varnothing_{inside}$ = 10 mm, Impex HighTech) allows to observe almost the complete sample volume ($\approx$ 99%). At the top, the screw driven piston, made of polyoxymethylene (POM), seals the cell via a sealing ring (6 x 2 mm, NBR 90) against the inner surface of the sapphire ring cylinder. Thereby the piston allows adjusting the cell volume and thus the pressure. At the bottom of the sapphire ring cylinder, another sealing ring (14 x 3 mm, NBR 90) seals to a block of brass. A small bore ($\varnothing$ = 0.3 mm) and a thin capillary connects the sample volume to a diaphragm (stainless steel, $\varnothing$ = 22 mm, 0.05 mm thick) and a needle valve (Nr. 8002-0120, LATEK). The diaphragm (sealed via two sealing rings: $\varnothing$ = 21.8-17 mm, 0.3 mm thick, POM) separates the sample volume from the miniature pressure transducer (Type 81530-500, maximum pressure 500 bar, Burster). The pressure transfer is ensured via an air-free water reservoir. The sapphire ring cylinder was designed to withstand pressures up to p = 800 bar and checked up to p = 450 bar. Though the pressure was only varied up to p = 300 bar. To control the temperature of the sample, the whole cell is placed in a thermostated water bath. Thus the temperature can be controlled in the range of $T \approx 3 - 90\ °C$ with a precision up to $\Delta T = \pm 0.1\ K$.

The samples were directly prepared inside the cell, while a magnetic stirring bar allows for the homogenization of the sample. First, at $p = 1$ bar and $T = 21 \pm 1\ °C$ liquid or solid components were weighed into the cell, with an accuracy of $\Delta m = \pm 0.0001\ g$. The cell was sealed with the piston and placed upside down in a home-built filling apparatus, which is used in order to fill in CO_2. As the cell is placed upside down, the valve can be used to fill in CO_2 from the top. The main component of the filling apparatus is a diaphragm accumulator (SBO 450 – 0.6 A6 / 342 U – 250 AK, HYDAC), which is filled in the first step with CO_2 from a dip-tube bottle. Then the CO_2 is filled in the cell and the pressure is increased by using N_2 ($p \approx 90$ bar). This allows filling in liquid CO_2 at a pressure of $p = 72$ bar and room temperature. The amount of CO_2 is determined by weighing the cell ($m \approx 4\ kg$) before and after filling with CO_2, with an accuracy of $\Delta m = \pm 0.01\ g$ (balance: MSE4202S, Sartorius). In this way, an accuracy of $\Delta m(CO_2) = \pm 0.014\ g$ for the mass of CO_2 $m(CO_2)$ results.

In order to study the temperature-dependent phase behavior the cell is placed in a temperature-controlled (thermostat DC30, Haake) water bath (Figure 6.3).

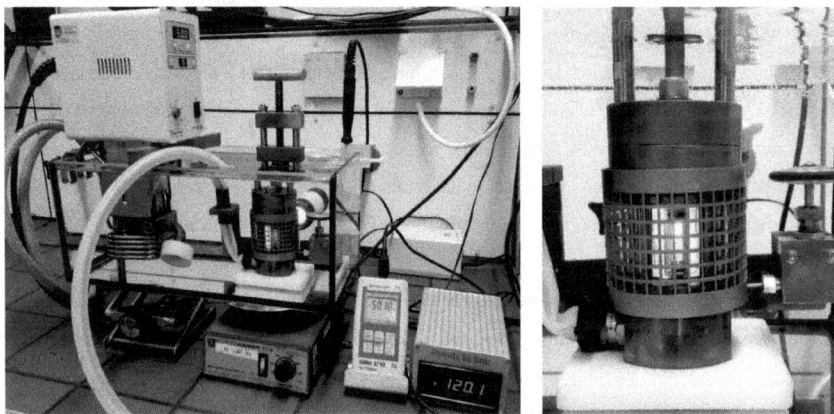

Figure 6.3: High-pressure view cell placed in the water bath above the stirring plate. The temperature is adjusted by the thermostat and controlled by the digital thermometer. The tunable piston allows adjusting the sample volume and thus the pressure with the external pressure display. The sapphire ring cylinder allows the visual inspection of the phase behavior using a microscope lamp and two crossed polarizers. For safety reasons, a protective cage is placed around the sapphire ring cylinder.

The temperature was measured by a digital thermometer (GMH 3710, $\Delta T = \pm 0.01$ K, GHM-Greisinger). After reaching the intended temperature equilibrium the pressure was adjusted. The type and number of coexisting phases were determined without stirring by means of visual inspection of transmitted and scattered light, using a microscope lamp (Belani, Gerhardt Optik). The occurrence of liquid crystalline phases was checked via two crossed polarizers. The temperature was varied until all phase boundaries for the envisaged pressures were determined (± 0.1 K).

6.3.2 Stroboscopic high-pressure SANS cell

All SANS-measurements at elevated pressure and the pressure jump measurements were performed using a home-built stroboscopic high-pressure SANS (SHP-SANS) view cell [22, 46, 59]. The cell was designed by the mechanical workshop of the Institute of Physical Chemistry, Cologne.

The cell can be filled with liquids as well as gases, while a metal bellow inside the cell enables the generation of periodic pressure cycles with adjustable amplitude and frequency for periods up to several hours. The core of the cell, made of bronze, is equipped with two parallel sapphire windows (thickness 12 mm, neutron path way 2 mm) allowing to check the phase behavior before and after the SANS-measurement, while an adequate transmission of neutrons is provided at the same time. The windows are stepped, i.e. have diameters of 24 and 18 mm (entrance) and 44 and 32 mm (exit), respectively. The windows are sealed by sealing rings (NBR 90) of 18 x 2 mm (entrance) and 32 x 2 mm (exit), respectively. Above and below the windows the sample volume (about 13 cm^3) becomes cylindrical ($\varnothing = 16$ mm). The setup of the SHP-SANS cell allows measurements in the range of $T = 5 - 70\ °C$ ($\Delta T = \pm 0.1$ K) and $p = 1 - 300$ bar ($\Delta p = \pm 5$ bar).

A mechanical (left) and 3D-rendered (right) cross section through the SHP-SANS cell is shown in Figure 6.4. The neutron beam (highlighted in red, right) enters the cell from the left via the smaller entrance window and is scattered by the sample (shown in orange, right). The larger exit window and the conical shape of the SHP-SANS cell (opening angle of 64°) on the right side allows for a full illumination of the detector even at shortest sample-to-detector distances. From the top, a screw driven piston closes the cell via a sealing ring (12 x 2 mm, NBR 90) at the inner surface of the cell. Moreover, the piston allows to adjust the cell volume and thus the pressure. For solely static measurements at constant pressures p a piston made of solid POM is used. For pressure jump, i.e. kinetic, measurements a hollow piston with an also hollow metal bellow (shown in green, right) is used instead. The bellow is filled with hydraulic oil, while a hydraulic circuit (shown in yellow, right) enables periodic elongation and shrinkage of the bellow. Since the bellow is inside the cell, this corresponds to periodic changes of the cell volume and therewith the pressure. At the bottom a metal block closes the cell volume via a sealing ring (18 x 3 mm, NBR 90). A small bore ($\emptyset = 0.3$ mm) and a thin capillary connects the sample volume to a diaphragm (stainless steel, $\emptyset = 22$ mm, 0.05 mm thick) and a needle valve (Nr. 8002-0121, LATEK). The diaphragm (sealed via two sealing rings: $\emptyset = 21.8$-17 mm, 0.3 mm thick, POM) separates the sample volume from the miniature pressure transducer (Type 81530-500, maximum pressure 500 bar, Burster). The pressure transfer is ensured via a capillary, which was air-free filled with water.

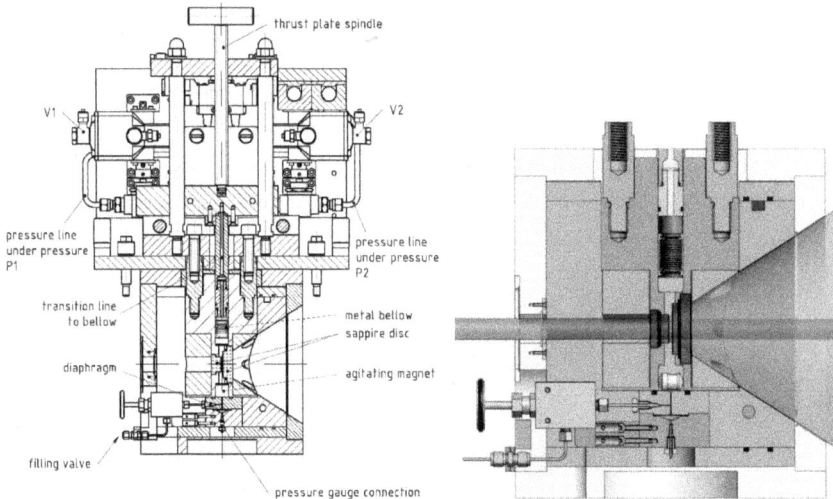

Figure 6.4: Mechanical cross section (left) and 3D-rendered cross section (right) through the SHP-SANS cell. The thrust plate spindle moves the piston, while the metal bellow (shown in green, right) and piston allows the variation of the cell volume (orange) and thus the pressure. The hydraulic circuit used for periodic pressure cycles is shown in yellow. The cell is equipped with two sapphire windows (thickness 12 mm) installed in parallel at a distance of 2 mm. The neutron beam (highlighted in red) enters on the left. The conical shape on the right, allows the scattering up to an angle of 64°. The cell volume $V \sim 13$ cm^3 is sealed via the piston and a diaphragm at the bottom.

A dumbbell-shape magnetic stirring bar placed inside the cell below the windows allows the sample to be homogenized. Note that the stirring bar is larger than the gap of the windows. A stirring plate (not shown in Figure 6.4) is directly mounted below the cell to enable an agitating of the stirring bar. The pressure is displayed and recorded by a PC Oscilloscope PicoScope 3204. The temperature is measured via a temperature sensor (PT100 A class, RS Components GmbH) close to the cell volume. The core of the cell is placed in a thermo-casing made of aluminium with meander-like water-channels (see [224]). By connecting the thermo-casing (insulated with POM) to a thermostat a temperature of $T = 5 - 70$ °C ($\Delta T = \pm 0.1$ K) is accessible. The tunable combination of piston and metal bellow allows varying the pressure within $p = 1 - 300$ bar ($\Delta p = \pm 5$ bar).

For kinetic measurements the hollow metal bellow is connected via a hollow connection piece, made of POM, to the hollow tunable piston (Figure 6.5). The connections are sealed via sealing rings (made of NBR 90) at the top (9 x 2 mm) and bottom (8 x 2 mm). The bellow/piston is filled with hydraulic oil, while the hydraulic circuit is connected to a hydraulic pump (Figure 6.6). Thereby the hydraulic pressure controls the volume and thus the length of the bellow inside the cell. Via two outlets the hydraulic pump provides two hydraulic flows, each with an adjustable pressure $p_1, p_2 < 500$ bar. Periodic elongation and shrinkage of the bellow corresponds to periodic changes of the cell volume and thus the pressure.

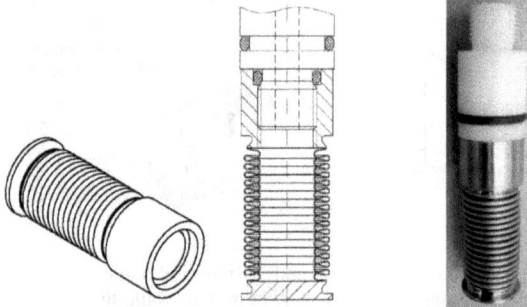

Figure 6.5: Technical drawings (left) and photograph (right) of the hollow metal bellow. Via a hollow connection piece, made of POM, the bellow is attached to an also hollow tunable piston. A sealing ring (12 x 2 mm, NBR 90) closes the cell volume from the top. The bellow is filled with hydraulic oil, whose pressure is controlled by a hydraulic pump. The hydraulic pressure controls the volume and thus the length of the bellow inside the cell. Periodic elongation and shrinkage of the bellow corresponds to periodic changes of the cell volume and therewith the pressure.

The pneumatic and hydraulic circuit diagram of the SHP-SANS cell is shown in Figure 6.6. Hydraulic ball valves (V1 and V2) control the connection of the two hydraulic flows to the piston/bellow. In order to increase the length of the bellow and thus the pressure of the sample, the valve V1 is opened and the hydraulic pressure $p_1 \sim p_{high}$ is transferred to the bellow. Subsequently, the valve is closed and a constant pressure obtained. Short-term opening of V2 transfers the hydraulic pressure $p_2 \sim p_{low}$ to the bellow and leads to a shrinking of the bellow and thus the sample pressure decreases. Periodic repetition in this way generates a periodic pressure cycle. V1 and V2 are operated by an antagonistic circuit of four pneumatic valves, while one pneumatic valve each is used to open or close one hydraulic valve. The pneumatic valves are controlled using compressed air (p = 8–10 bar). In this way, switching times of up to 11 ms could be obtained for the hydraulic ball valves [59]. Hence, the maximum adjustable frequency amounts to $v < 23$ Hz. A pressure cycle consists of four steps with the respective duration t: V1 open – V1 closed – V2 open – V2 closed. Note, that the pressure cycle can be repeated as often as necessary.

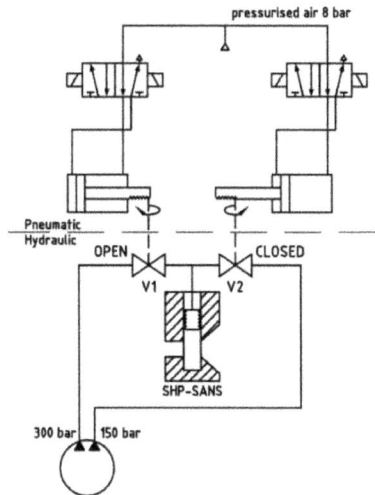

Figure 6.6: Pneumatic and hydraulic circuit diagram of the SHP-SANS cell. Via two outlets the hydraulic pump (circle) provides two adjustable pressures $p_1, p_2 < 500$ bar. The valves V1 and V2 control the connection of the two flows to the piston/bellow and thus the pressure within the cell. V1 and V2 are operated by an antagonistic pneumatic circuit with pressurized air at p = 8–10 bar. One pneumatic valve each to open or close one hydraulic valve enables the generation of pressure cycles with adjustable frequency of $v < 23$ Hz.

To avoid overheating, we use a thermostat and two fans to cool the hydraulic pump. The periodic pressure cycle can be adjusted by the hydraulic pressures p_1 and p_2 at the hydraulic pump. The switching times of the pneumatic valves are operated by a LABVIEW-based software program displaying also the values of p_1 and p_2. Furthermore, the program transmits a trigger signal to the D11 instrument. Having defined and started a kinetic count, the detector starts counting upon the trigger signal. Additional 0.1 s within the pressure cycle were used to ensure the data transfer.

All samples were directly prepared in the cell. First, at ambient conditions liquid or solid components were weighed ($\Delta m = \pm 0.0001$ g) into the cell. The cell was sealed with the respective piston and turned upside down using a home-built holding. Liquid CO_2 was filled in at $T = 23$ °C and $p = 72$ bar via a home-built filling apparatus and a diaphragm accumulator (SBO 250 – 2 A6 / 346 A – 180 AK, HYDAC) [224]. The position of the tunable piston was therefore accurately determined using a dial gauge. Knowing the exact volume of the cell and the volumes of the solid and liquid components, the amount of CO_2 was determined. The sample is homogenized by the stirring bar and by turning the cell upside down. The phase behavior is checked or determined prior to the SANS-measurements (see procedure in 6.3.1). Subsequently, the SHP-SANS cell was placed on the sample mount of the D11 or D22 instrument via an adapter plate. Note that for all experiments an aperture with a circular opening of $\varnothing$ = 13 mm was used in front of the SHP-SANS cell. The instrumental parameters are summarized in Table 6.8.

6.3.3 Small Angle Neutron Scattering at ambient pressure

SANS-samples at ambient pressure were kept in Hellma 404-QX cuvettes (Hellma) with a neutron path way of 1 mm. A home-built cell holder with a very high temperature stability of $\Delta T = \pm 0.03$ K was used to maintain a constant temperature. For more details see [151]. The temperature is controlled via a thermostat, as with the SHP-SANS cell. Through a bajonette-type fitting the cell holder can be mounted on the adapter plate, which is also used for the SHP-SANS cell. Note that for all experiments an aperture with a circular opening of $\varnothing$ = 13 mm was used in front of the cell holder. The instrumental parameters are summarized in Table 6.9.

6.3.4 Neutron Spin Echo

All neutron spin echo measurements were performed at the IN15 spectrometer at the Institute Laue-Langevin (ILL) in Grenoble, France. The measurements were carried out using the home-built stroboscopic high-pressure SANS cell (see 6.3.2). The cell is made completely out of non-magnetic substances such as bronze. Screws from stainless steel were exchanged by demagnetized ones or simply replaced by another non-magnetic material. The stirring plate was dismounted and the stirring bar removed prior to the NSE beam time. To allow for a homogenization of the microemulsion sample glass beads ($\varnothing$ = 1.25–1.65 mm, Roth) of known volume were inserted into the cell. By repeatedly turning the cell upside down the samples are mixed in this way. Homogeneity of the sample was ensured by visual inspection through the sapphire windows. The cell was placed on the sample position via an adapter plate. The resolution of the spin-echo spectrometer was determined using a graphite plate directly behind the sapphire window on the detector side respectively directly at the cell position without the cell. The NSE spectra were recorded at three different detector angles (3.5, 6.5 and 9.5°) thus at 15 different q values of q = 0.0273 Å^{-1} to 0.1182 Å^{-1} and a wavelength of λ = 10 Å ($\Delta\lambda/\lambda$ = 15%, full width at half maximum). Raw data evaluation was done using the software available at IN15.

Moreover, the miniature pressure transducer was connected to the IN15 instrument and the pressure incorporated in the software program. Thus, for each *Fourier* time τ the current pressure was saved in the result file.

6.3.5 Surface tension

In order to characterize the static surface tension of water or polyol as a function of the volume fraction of surfactant the Du Nuoy ring method [219] was applied. Therefore, a ring tensiometer (STA-1, SINTERFACE) [225] in combination with an electronic microbalance (CP64, Sartorius) was used. Thus, the weight of the liquid meniscus formed by a thin platinum wire with a radius of R_r = 9.55 mm is determined. By an automatically motor driven movement of the ring the maximum force F_{max} is measured as a function of the time. Depending on the ring geometry and the sample density a correction factor f_{corr} is required. f_{corr} was determined according to Harkins and Jordan [226]. Consequently, the surface tension σ_a can be determined as $\sigma_a = F_{max}/(4\pi R_r f_{corr})$. The determination of the maximum force was carried out every 100 seconds, while the measurements were performed over a period of about one hour.

6.3.6 Density measurements

A density oscillator (DMA^{TM} 5000 M, Anton Paar) [227] was used in order to determine the density of a liquid sample. A hollow U-shaped glass tube is filled with the sample (approximately 1 ml). The tube is electronically excited to undamped oscillations. The resulting eigenfrequency is related to the oscillating mass. With known volume, the mass of the sample and thus the density can be determined within a temperature range of $T = 0 - 100$ °C and an accuracy of $\Delta T = \pm 0.01$ K.

The density of solid components was determined using a pycnometer (Hecht) of certified volume $V = 10$ cm^3 and attachable thermometer ($\Delta T = \pm 0.2$ K). At the temperature T the macroscopic density ρ_i of the substance i is given according to

$$\rho_i = \frac{m_2 - m_0}{(m_1 - m_0) - (m_3 - m_2)} \rho_{H_2O}$$

　　　　　　　　　　　　　　　　　　　　　　　　　　　　　　6.1

With m_0 being the mass of the empty pycnometer, m_1 the mass of the pycnometer filled with water, m_2 the mass of the pycnometer with the substance ($m_2 - m_0 \leq 2$ g), m_3 the mass of the pycnometer filled with the substance and water and ρ_{H_2O} the known macroscopic density of water. Note that the substances were melted and cooled again before the pycnometer was filled with water.

6.4 Tables

6.4.1 SANS-samples

Table 6.2: SANS $T(w_B)$-samples $D_2O/NaClO_4 - CO_2 - FS\text{-}3100$ ($\gamma_{a,p} = 0.08$, $\varepsilon_p = 0.05$) measured at D11. Note that m_{CO_2} is the target weight, adjusted by the position of the piston at 33.70 mm.

Sample	$m_{D_2O/NaClO_4}$ / g	$m_{FS\text{-}3100}$ / g	m_{CO_2} / g	$\gamma_{a,p}$	$w_{B,p}$
SL10	9.860	0.7790	1.26	0.080	0.116

Table 6.3: SANS $T(\gamma)$-samples $D_2O/NaClO_4 - CO_2 - FS\text{-}3100$ ($\alpha_p = 0.40$, $\varepsilon_p = 0.05$) measured at D11. Note that m_{CO_2} is the target weight, adjusted by the position of the piston at 33.70 mm.

Sample	$m_{D_2O/NaClO_4}$ / g	$m_{cyclohexane}$ / g	$m_{FS\text{-}3100}$ / g	m_{CO_2} / g	$\gamma_{a,p}$	$w_{B,p}$	β	γ_p
SL1	5.1737	-	3.3413	3.10	0.417	0.280	0.00	0.30
SL2	5.1704	0.7306	3.3375	2.37	0.417	0.280	0.20	0.30
SL3	5.1744	0.2600	3.3370	2.84	0.417	0.280	0.08	0.30
SL4	5.1719	0.5162	3.3383	2.59	0.417	0.280	0.15	0.30
SL5	5.6905	-	2.1435	3.42	0.294	0.320	0.00	0.20
SL6	5.6932	0.8094	2.1402	2.61	0.294	0.320	0.20	0.20

Table 6.4: SANS $T(w_A)$-samples $D_2O/NaClO_4 - CO_2 - FS\text{-}3100$ ($\gamma_b = 0.25$, $\varepsilon_p = 0.05$) measured at D11. Note that m_{CO_2} is the target weight, adjusted by the position of the piston at 33.70 mm.

Sample	$m_{D_2O/NaClO_4}$ / g	$m_{FS\text{-}3100}$ / g	m_{CO_2} / g	$\gamma_{a,p}$	$w_{B,p}$	$w_{A,p}$
SL7	0.8128	2.2656	6.79	0.755	0.694	0.075
SL8	0.9736	2.2326	6.69	0.716	0.683	0.090
SL9	1.1470	2.1981	6.59	0.681	0.671	0.105

Table 6.5: Compositions of the pseudo-binary SANS samples of the protonated PES NFZ 6509/glycerol – Q2-5211/Genapol® T250p ($\psi_{NFZ\ 6509} = 0.946$, $\psi_{glycerol} = 0.054$, $\delta = 0.50$) and partly deuterated d-PES NFZ 6509/glycerol-d_8 – Q2-5211/Genapol® T250p ($\psi_{d-NFZ\ 6509} = 0.942$, $\psi_{glycerol-d_8} = 0.058$, $\delta = 0.50$) system measured at D11. Note that for both the protonated and the partly deuterated polyol-mixture the molar ratio was kept constant.

Sample	type	m_{polyol} / g	$m_{Q2-5211}$ / g	$m_{Genapol®}$ / g	$\gamma_{a,p}$	$\gamma_{a,d}$
SL_1_PO	protonated	1.68917	0.21816	0.21843	0.205	-
SL_2_PO	partly deuterated	1.25580	0.15652	0.15652	0.203	0.200
SL_3_PO	protonated	1.93581	0.06098	0.06105	0.059	-
SL_4_PO	partly deuterated	1.44729	0.04456	0.04476	0.059	0.058

Table 6.6: Compositions of the bicontinuous SANS samples of the ternary system PES NFZ 6509/glycerol – CO_2/AAA – Q2-5211/Genapol® T250p ($\psi_{NFZ\ 6509} = 0.946$, $\psi_{glycerol} = 0.054$, $\delta = 0.50$) measured at D11 with 1,2-decanediol as AAA. Note that, except for SL1p, the deuterated polyol mixture d-PES NFZ 6509/glycerol-d_8 ($\psi_{d-NFZ\ 6509} = 0.942$, $\psi_{glycerol-d_8} = 0.058$) was used, while the molar ratio was kept constant. The composition parameters were calculated from m_{CO_2}, which is the target weight, adjusted by the position of the piston at 33.70 mm. SL1p, SL2p and SL3p were measured using the static piston of the SHP-SANS cell, while SL4p and SL5p were measured using the kinetic piston.

Sample	m_{polyol} / g	$m_{Q2-5211}$ / g	$m_{Genapol®}$ / g	m_{AAA} / g	m_{CO_2} / g	γ_a	w_B	α (α_{CO_2})	β	γ
SL1p	8.639	1.1413	1.1428	-	1.47	0.21	0.12	0.15	-	0.18
SL2p	8.782	1.1142	1.1146	-	1.51	0.21	0.12	0.15	-	0.18
SL3p	8.127	1.4706	1.4706	-	1.38	0.27	0.11	0.15	-	0.24
SL4p	10.397	1.3168	1.3167	-	1.80	0.21	0.12	0.15	-	0.18
SL5p	10.080	1.3153	1.3153	0.3165	1.74	0.21	0.14	0.17 (0.15)	0.15	0.18

6.4.2 NSE-samples

Table 6.7: Compositions of the bicontinuous NSE samples of the microemulsion system $D_2O/NaClO_4 - CO_2/cyclohexane - FS-3100$ ($\alpha_p = 0.40$, $\varepsilon_p = 0.05$) measured at IN15. Note that m_{CO_2} is the target weight, adjusted by the position of the piston at 33.70 mm.

Sample	$m_{D_2O/NaClO_4}$ / g	$m_{cyclohexane}$ / g	$m_{FS-3100}$ / g	$m_{glassbeads}$ / g	m_{CO_2} / g	$\gamma_{a,p}$	$w_{B,p}$	β	γ_p
SL1	5.1715	-	3.3386	2.5010	3.10	0.417	0.280	0.00	0.30
SL2	5.1719	0.7374	3.3385	2.5004	2.36	0.417	0.280	0.20	0.30
SL3	5.1684	0.2598	3.3407	2.4993	2.84	0.417	0.280	0.08	0.30
SL4	5.1692	0.5227	3.3370	2.4996	2.58	0.417	0.280	0.15	0.30
SL5	5.6888	-	2.1410	2.5004	3.42	0.294	0.320	0.00	0.20
SL6	5.6920	0.8045	2.1408	2.5009	2.61	0.294	0.320	0.20	0.20

6.4.3 Instrumental parameters

Table 6.8: Overview of the instrumental parameters for the SANS-measurements at elevated pressure using the SHP-SANS cell. The detector distances, collimation lengths and attenuator for the transmission are in brackets.

sample	instrument	λ / Å	detector distance / m	collimation length / m	attenuator
polyol (PES), $w_{A,p} = 0.105$, $w_{B,p} = 0.116$	D11	4.6	1.4 8 39 (8)	20.5 20.5 40.5 (8)	-- -- -- (3)
polyol (VP.PU)	D22	6	3 17.6 (11.2)	11.2 17.6 (11.2)	-- -- (500)
aqueous $T(\gamma)$-samples, $w_{A,p} = 0.075$, $w_{A,p} = 0.09$	D11	6	1.75 10 31 (10)	10.5/20.5 20.5 31 (10.5)	-- -- -- (2)

Table 6.9: Overview of the instrumental parameters for the SANS-measurements at ambient pressure using Hellma 404-QX cuvettes with 1 mm neutron pathway. The detector distances, collimation lengths and attenuator for the transmission are in brackets.

sample	instrument	λ / Å	detector distance / m	collimation length / m	attenuator
polyol (PES)	D11	4.6	1.4 8 (8)	20.5 20.5 (8)	-- -- (3)
polyol (VP.PU)	D22	6	3 17.6 (11.2)	11.2 17.6 (17.6)	-- -- (500)

7 References

[1]   V. Bergeron and P. Walstra, Foams, in *Fundamentals of Interface and Colloid Science*, ed. J. Lyklema, Elsevier Ltd., **2005**, vol. *5*, 1–38.

[2]   A. Demharter, "Polyurethane rigid foam, a proven thermal insulating material for applications between +130°C and -196°C", *Cryogenics* **1998**, *38*, 113–117.

[3]   N. Adam, G. Avar, H. Blankenheim, W. Friederichs, M. Giersig, E. Weigand, M. Halfmann, F.-W. Wittbecker, D.-R. Larimer, U. Maier, S. Meyer-Ahrens, K.-L. Noble and H.-G. Wussow, Polyurethanes, in *Ullmann's Encyclopedia of Industrial Chemistry*, Wiley-VCH Verlag GmbH & Co. KGaA, **2000**, 10.1002/14356007.a21_665.pub2.

[4]   S. S. Kistler, "Coherent expanded aerogels and jellies", *Nature (London, U. K.)* **1931**, *127*, 741.

[5]   S. S. Kistler, "Coherent expanded aerogels", *J. Phys. Chem.* **1932**, *36*, 52–64.

[6]   B. Notario, J. Pinto and M. A. Rodriguez-Perez, "Nanoporous polymeric materials: A new class of materials with enhanced properties", *Prog. Mater Sci.* **2016**, *78–79*, 93–139.

[7]   D. Miller and V. Kumar, "Microcellular and nanocellular solid-state polyetherimide (PEI) foams using sub-critical carbon dioxide II. Tensile and impact properties", *Polymer* **2011**, *52*, 2910–2919.

[8]   J. Fricke, "FCKW – freie Wärmedämmungen", *Physik in unserer Zeit* **1989**, *6*.

[9]   E. Placido, M. C. Arduini-Schuster and J. Kuhn, "Thermal properties predictive model for insulating foams", *Infrared Physics & Technology* **2005**, *46*, 219–231.

[10]   S. H. Singh, Blowing agents for polyurethane foams, in *Rapra Rev. Rep.*, Rapra Technology Ltd., Shawbury, U.K., **2002**, vol. *12*, 160.

[11]   H. Weber, I. De Grave, E. Röhrl and V. Altstädt, Foamed Plastics, in *Ullmann's Encyclopedia of Industrial Chemistry*, Wiley-VCH Verlag GmbH & Co. KGaA, **2000**, 10.1002/14356007.a11_435.pub2.

[12]   R. Baetens, B. P. Jelle, J. V. Thue, M. J. Tenpierik, S. Grynning, S. Uvsløkk and A. Gustavsen, "Vacuum insulation panels for building applications: A review and beyond", *Energy and Buildings* **2010**, *42*, 147–172.

[13]   M. Knudsen, "Eine Revision der Gleichgewichtsbedingung der Gase. Thermische Molekularströmung", *Annalen der Physik* **1909**, *336*, 205–229.

[14]   X. Lu, R. Caps, J. Fricke, C. T. Alviso and R. W. Pekala, "Correlation between structure and thermal conductivity of organic aerogels", *J. Non-Cryst. Solids* **1995**, *188*, 226–234.

[15] G. Reichenauer, U. Heinemann and H. P. Ebert, "Relationship between pore size and the gas pressure dependence of the gaseous thermal conductivity", *Colloids and Surfaces A: Physicochemical and Engineering Aspects* **2007**, *300*, 204–210.

[16] N. Leventis, P. Vassilaras, E. F. Fabrizio and A. Dass, "Polymer nanoencapsulated rare earth aerogels: chemically complex but stoichiometrically similar core-shell superstructures with skeletal properties of pure compounds", *J. Mater. Chem.* **2007**, *17*, 1502–1508.

[17] B. C. Tappan, M. H. Huynh, M. A. Hiskey, D. E. Chavez, E. P. Luther, J. T. Mang and S. F. Son, "Ultralow-Density Nanostructured Metal Foams: Combustion Synthesis, Morphology, and Composition", *Journal of the American Chemical Society* **2006**, *128*, 6589–6594.

[18] M. B. Bryning, D. E. Milkie, M. F. Islam, L. A. Hough, J. M. Kikkawa and A. G. Yodh, "Carbon nanotube aerogels", *Adv. Mater. (Weinheim, Ger.)* **2007**, *19*, 661–664.

[19] N. Husing and U. Schubert, "Aerogels – airy materials: chemistry, structure, and properties", *Angew. Chem., Int. Ed.* **1998**, *37*, 22–45.

[20] M. J. van Bommel and A. B. de Haan, "Drying of silica gels with supercritical carbon dioxide", *Journal of Materials Science* **1994**, *29*, 943–948.

[21] R. Strey and A. Müller, *Generation of nanodisperse inclusions in a high-viscosity matrix*, WO2011EP71541 20111201, WO2012072755A2, **2012**.

[22] A. Müller, *Preparation of Polymer Nano-Foams: Templates, Challenges and Kinetics*, PhD-Thesis, University of Cologne, **2013**, Cologne.

[23] S. A. E. Boyer and J. P. E. Grolier, "Modification of the glass transitions of polymers by high-pressure gas solubility", *Pure Appl. Chem.* **2005**, *77*, 593–603.

[24] L. Grassberger, *Towards cost-efficient preparation of nanoporous materials: formation kinetics, process optimization and material characterization*, PhD-Thesis, University of Cologne, **2017**, Cologne.

[25] L. Grassberger, K. Koch, R. Oberhoffer, A. Müller, H. F. M. Klemmer and R. Strey, "Blowing agent free generation of nanoporous poly(methylmethacrylate) materials", *Colloid & Polymer Sci* **2017**, *2*, 379–389.

[26] L. Grassberger and R. Strey, Experimental confirmation of the Knudsen effect in nanoporous insulation materials, presented in part at the ICNAA, Helsinki, **2017**.

[27] M. Schwan, T. Sottmann and R. Strey, *Aufgeschäumtes Material und Herstellverfahren für das aufgeschäumte Material, DE Pat.*, DE 102 60 815 A1, **2003**.

[28] M. Schwan, L. G. A. Kramer, T. Sottmann and R. Strey, "Phase behavior of propane- and scCO$_2$-microemulsions and their prominent role for the recently proposed foaming

procedure POSME (Principle of Supercritical Microemulsion Expansion)", *Phys. Chem. Chem. Phys.* **2010**, *12*, 6247–6252.

[29]   NIST Standard Reference Data, 03/01/2018, http://webbook.nist.gov/chemistry/fluid/.

[30]   M. Klostermann, *Quantitative Characterization of Supercritical Microemulsions and their Use for Preparing Highly Porous Sugar Foams*, PhD-Thesis, University of Cologne, **2011**, Cologne.

[31]   R. Strey, M. Klostermann, L. G. A. Kramer, R. Schwering and T. Sottmann, *Sugar Micro/Nanofoams, De Pat.*, US200913054138 20090716, **2011**.

[32]   G. Roshan Deen, C. L. P. Oliveira and J. S. Pedersen, "Phase behavior and kinetics of phase separation of a nonionic microemulsion of $C_{12}E_5$/water/1-chlorotetradecane upon a temperature quench", *The Journal of Physical Chemistry B* **2009**, *113*, 7138–7146.

[33]   S. Egelhaaf, U. Olsson, P. Schurtenberger, J. Morris and H. Wennerstrom, "Quantitative measurements of Ostwald ripening using time-resolved small-angle neutron scattering", *Physical Review E* **1999**, *60*, 5681–5684.

[34]   W. I. Higuchi and J. Misra, "Physical Degradation of Emulsions Via the Molecular Diffusion Route and the Possible Prevention Thereof", *J. Pharm. Sci.* **1962**, *51*, 459-&.

[35]   W. Koch and S. K. Friedlander, "Particle growth by coalescence and agglomeration", *Particle & Particle Systems Characterization* **1991**, *8*, 86–89.

[36]   P. Meakin, "Steady state droplet coalescence", *Physica A: Statistical Mechanics and its Applications* **1991**, *171*, 1–18.

[37]   E. Khazova, *Polymer-Nanoschäume aus Mikroemulsionen*, PhD-Thesis, University of Cologne, **2010**, Cologne.

[38]   D. Engelen, *Selbststrukturierte Systeme zur Bildung von Polyurethan-Nanoschäumen*, PhD-Thesis, University of Cologne, **2013**, Cologne.

[39]   S. Lindner, W. Friederichs, R. Strey, E. Khazova, T. Sottmann, D. Engelen and A. Chalbi, *Verfahren zur Herstellung eines geschäumten Materials, hierbei eingesetzte emulsionsförmige Zusammensetzung und hieraus erhältliches geschäumtes Material*, DE20101060386 20101105, **2012**.

[40]   J. D. Holmes, D. C. Steytler, G. D. Rees and B. H. Robinson, "Bioconversions in a water-in-CO2 microemulsion", *Langmuir* **1998**, *14*, 6371–6376.

[41]   M. Klostermann, R. Strey, T. Sottmann, R. Schweins, P. Lindner, O. Holderer, M. Monkenbusch and D. Richter, "Structure and dynamics of balanced supercritical CO_2-microemulsions", *Soft Matter* **2012**, *8*, 797–807.

[42]   A. Iezzi, P. Bendale, R. M. Enick, M. Turberg and J. Brady, "'Gel' formation in carbon dioxide-semifluorinated alkane mixtures and phase equilibria of a carbon dioxide-perfluorinated alkane mixture", *Fluid Phase Equilibria* **1989**, *52*, 307–317.

[43]   K. A. Consani and R. D. Smith, "Observations on the solubility of surfactants and related molecules in carbon dioxide at 50°C", *The Journal of Supercritical Fluids* **1990**, *3*, 51–65.

[44]   C. M. Butt, D. C. G. Muir and S. A. Mabury, "Biotransformation pathways of fluorotelomer-based polyfluoroalkyl substances: A review", *Environ. Toxicol. Chem.* **2014**, *33*, 243–267.

[45]   M. Klostermann, T. Foster, R. Schweins, P. Lindner, O. Glatter, R. Strey and T. Sottmann, "Microstructure of supercritical CO_2-in-water microemulsions: a systematic contrast variation study", *Phys. Chem. Chem. Phys.* **2011**, *13*, 20289–20301.

[46]   A. Müller, Y. Puetz, R. Oberhoffer, N. Becker, R. Strey, A. Wiedenmann and T. Sottmann, "Kinetics of pressure induced structural changes in super- or near-critical CO_2-microemulsions", *Phys. Chem. Chem. Phys.* **2014**, *16*, 18092–18097.

[47]   Y. Pütz, L. Grassberger, P. Lindner, R. Schweins, R. Strey and T. Sottmann, "Unexpected efficiency boosting in CO_2-microemulsions: a cyclohexane depletion zone near the fluorinated surfactants evidenced by a systematic SANS contrast variation study", *Phys. Chem. Chem. Phys.* **2015**, *17*, 6122–6134.

[48]   J. Eastoe, Z. Bayazit, S. Martel, D. C. Steytler and R. K. Heenan, "Droplet structure in a water-in-CO_2 microemulsion", *Langmuir* **1996**, *12*, 1423-1424.

[49]   J. Eastoe, B. M. H. Cazelles, D. C. Steytler, J. D. Holmes, A. R. Pitt, T. J. Wear and R. K. Heenan, "Water-in-CO_2 microemulsions studied by small-angle neutron scattering", *Langmuir* **1997**, *13*, 6980–6984.

[50]   T. Foster, T. Sottmann, R. Schweins and R. Strey, "Small-angle-neutron-scattering from giant water-in-oil microemulsion droplets. II. Polymer-decorated droplets in a quaternary system", *Journal of Chemical Physics* **2008**, *128*.

[51]   T. Foster, T. Sottmann, R. Schweins and R. Strey, "Small-angle neutron scattering from giant water-in-oil microemulsion droplets. I. Ternary system", *Journal of Chemical Physics* **2008**, *128*.

[52]   J. D. van der Waals, "Untersuchung über die übereinstimmenden Eigenschaften der Normallinien des gesättigten Dampfes und der Flüssigkeit", *Annalen der Physik Beiblätter* **1881**, *5*, 27–28.

[53]   T. Sottmann and R. Strey, "Evidence of corresponding states in ternary microemulsions of water-alkane-C_iE_j", *Journal of Physics-Condensed Matter* **1996**, *8*, A39–A48.

[54]   R. Strey, "Microemulsion microstructure and interfacial curvature", *Colloid Polym. Sci.* **1994**, *272*, 1005–1019.

[55]   T. Sottmann, *Mikroemulsionen: Eigenschaften von internen Grenzflächen*, Phd-Thesis, Georg-August-Universität Göttingen, **1997**, Göttingen.

[56]   T. Sottmann and R. Strey, Microemulsions, in *Fundamentals of Interface and Colloid Science*, ed. J. Lyklema, Elsevier Ltd., **2005**, vol. *5*, 1–96.

[57]   Y. Pütz, *Influence of hydrophobic additives on supercritical CO_2-microemulsions and their foaming properties*, Masterthesis, University of Cologne, **2012**, Cologne.

[58]   S. H. Chen, S. M. Choi and P. LoNostro, "Measurement of interfacial curvatures in micro-phase-separated bicontinuous structures using small-angle neutron scattering", *Nuovo Cimento Della Societa Italiana Di Fisica D-Condensed Matter Atomic Molecular and Chemical Physics Fluids Plasmas Biophysics* **1998**, *20*, 1971–1988.

[59]   Y. Pütz, *CO_2-microemulsions with additives Phase behaviour, microstructure and pressure-induced kinetics*, PhD-Thesis, University of Cologne, **2015**, Cologne.

[60]   M. Kahlweit and R. Strey, "Phae Behavior of Ternary Systems of the Type H_2O-Oil-Nonionic Amphiphile (Microemulsions)", *Angewandte Chemie-International Edition in English* **1985**, *24*, 654–668.

[61]   L. Prince, *Microemulsions: Theory and Practice*, Academic, **1977**.

[62]   T. P. Hoar and J. H. Schulman, "Transparent water-in-oil dispersions the oleopathic hydro-micelle", *Nature* **1943**, *152*, 102–103.

[63]   J. H. Schulman, W. Stoeckenius and L. M. Prince, "MECHANISM OF FORMATION AND STRUCTURE OF MICRO EMULSIONS BY ELECTRON MICROSCOPY", *Journal of Physical Chemistry* **1959**, *63*, 1677–1680.

[64]   H. Kunieda and S. E. Friberg, "Critical phenomena in a surfactant/water/oil system. Basic study on the correlation between solubilization, microemulsion, and ultralow interfacial tensions", *Bull. Chem. Soc. Jpn.* **1981**, *54*, 1010–1014.

[65]   H. Kunieda and K. Shinoda, "Phase behavior in systems of nonionic surfactant/water/oil around the hydrophile-lipophile-balance temperature (HLB temperature)", *J. Dispersion Sci. Technol.* **1982**, *3*, 233-244.

[66]   R. W. Gale, J. L. Fulton and R. D. Smith, "Organized Molecular Assemblies in the Gas Phase: Reverse Micelles and Microemulsions in Supercritical Fluids", *Journal of the American Chemical Society* **1987**, *109*, 920–921.

[67]   M. Kahlweit, R. Strey, D. Haase, H. Kunieda, T. Schmeling, B. Faulhaber, M. Borkovec, H. F. Eicke, G. Busse and a. et, "How to study microemulsions", *J. Colloid Interface Sci.* **1987**, *118*, 436–453.

[68]   G. J. T. Tiddy, "Surfactant-water liquid crystal phases", *Physics reports* **1980**, *57*, 1–46.

[69]   D. J. Mitchell, G. J. T. Tiddy, L. Waring, T. Bostock and M. P. McDonald, "Phase behaviour of polyoxyethylene surfactants with water. Mesophase structures and partial miscibility (cloud points)", *Journal of the Chemical Society, Faraday Transactions 1: Physical Chemistry in Condensed Phases* **1983**, *79*, 975–1000.

[70]   S. Burauer, T. Sachert, T. Sottmann and R. Strey, "On microemulsion phase behavior and the monomeric solubility of surfactant", *Phys. Chem. Chem. Phys.* **1999**, *1*, 4299–4306.

[71]   M. Kahlweit, R. Strey and G. Busse, "Microemulsions: a qualitative thermodynamic approach", *J. Phys. Chem.* **1990**, *94*, 3881–3894.

[72]   J. Brunner-Popela, R. Mittelbach, R. Strey, K. V. Schubert, E. W. Kaler and O. Glatter, "Small-angle scattering of interacting particles. III. D_2O-$C_{12}E_5$ mixtures and microemulsions with *n*-octane", *Journal of Chemical Physics* **1999**, *110*, 10623-10632.

[73]   M. Kahlweit, R. Strey and G. Busse, "Weakly to strongly structured mixtures", *Phys. Rev. E: Stat. Phys., Plasmas, Fluids, Relat. Interdiscip. Top.* **1993**, *47*, 4197–4209.

[74]   S. A. Safran, "Saddle-splay modulus and the stability of spherical microemulsions", *Physical Review A* **1991**, *43*, 2903-2904.

[75]   S. A. Safran and L. A. Turkevich, "Phase Diagrams for Microemulsions", *Physical Review Letters* **1983**, *50*, 1930–1933.

[76]   S. Burauer, *Elektronenmikroskopie komplexer Fluide*, Dissertation, Universität zu Köln, **2001**, Köln.

[77]   B. Jakobs, *Amphiphile Blockcopolymere als "Efficiency Booster" für Tenside: Entdeckung und Aufklärung des Effekts*, PhD-Thesis, University of Cologne, **2001**, Cologne.

[78]   B. Jakobs, T. Sottmann and R. Strey, "Efficiency boosting with amphiphilic block copolymers. A new approach to microemulsion formulation", *Tenside, Surfactants, Deterg.* **2000**, *37*, 357–364.

[79]   T. Sottmann and C. Stubenrauch, "Phase behaviour, interfacial tension and microstructure of microemulsions", *Microemulsions: Background, New Concepts, Applications, Perspectives* **2009**, 1–47.

[80]   P. G. de Gennes and C. Taupin, "Microemulsions and the flexibility of oil/water interfaces", *The Journal of Physical Chemistry* **1982**, *86*, 2294–2304.

[81]   J. Eastoe and J. S. Dalton, "Dynamic surface tension and adsorption mechanisms of surfactants at the air–water interface", *Adv. Colloid Interface Sci.* **2000**, *85*, 103-144.

[82]   J. J. Jasper, "The surface tension of pure liquid compounds", *Journal of Physical and Chemical Reference Data* **1972**, *1*, 841–1010.

[83]　J. W. Gibbs, "The collected works of J. Willard Gibbs. 1928", *New York: Longmans* **1928**.

[84]　R. Aveyard, B. P. Binks and P. D. I. Fletcher, Surfactant molecular geometry within planar and curved monolayers in relation to microemulsion phase behaviour, in *The Structure, Dynamics and Equilibrium Properties of Colloidal Systems*, Springer, **1990**, 557–581.

[85]　M. J. Rosen and D. S. Murphy, "Effect of the nonaqueous phase on interfacial properties of surfactants. 2. Individual and mixed nonionic surfactants in hydrocarbon/water systems", *Langmuir* **1991**, *7*, 2630–2635.

[86]　T. Sottmann, R. Strey and S. H. Chen, "A small-angle neutron scattering study of nonionic surfactant molecules at the water-oil interface: Area per molecule, microemulsion domain size, and rigidity", *Journal of Chemical Physics* **1997**, *106*, 6483–6491.

[87]　B. von Szyszkowski, "Experimentelle Studien über kapillare Eigenschaften der wässerigen Lösungen von Fettsäuren", *Z. Phys. Chem.* **1908**, *64*, 385–414.

[88]　S. Wurtz, *Polyol-Tensid-Systeme zur Entwicklung von neuartigen Polyurethan-Schäumen – Untersuchung der Grenzflächeneigenschaften und CO_2-Solubilisierung –*, GymPO I thesis (unpublished), Universität Stuttgart, **2016**.

[89]　D. F. Evans and H. Wennerstrom, *The Colloidal Domain: Where Physics, Chemistry, Biology, and Technology Meet*, Wiley VCH, 2 edn., **1999**.

[90]　R. Strey, W. Jahn, G. Porte and P. Bassereau, "Freeze fracture electron microscopy of dilute lamellar and anomalous isotropic (L3) phases", *Langmuir* **1990**, *6*, 1635–1639.

[91]　W. Helfrich, "Elastic Properties of Lipid Bilayers: Theory and Possible Experiments", *Zeitschrift Fur Naturforschung C-a Journal of Biosciences* **1973**, *C 28*, 693–703.

[92]　D. C. Morse, "Topological instabilities and phase behavior of fluid membranes", *Physical Review E* **1994**, *50*, R2423–R2426.

[93]　G. Gompper and D. M. Kroll, "Membranes with fluctuating topology: Monte Carlo simulations", *Physical Review Letters* **1998**, *81*, 2284–2287.

[94]　B. Farago, D. Richter, J. S. Huang, S. A. Safran and S. T. Milner, "Shape and Size Fluctuations of Microemulsion Droplets: The Role of Cosurfactant", *Physical Review Letters* **1990**, *65*, 3348–3351.

[95]　J. S. Huang, S. T. Milner, B. Farago and D. Richter, "Study of Dynamics of Microemulsion Droplets by Neutron Spin-Echo Spectroscopy", *Physical Review Letters* **1987**, *59*, 2600–2603.

[96]   T. Hellweg and D. Langevin, "Bending elasticity of the surfactant monolayer in droplet microemulsions: Determination by a combination of dynamic light scattering and neutron spin-echo spectroscopy", *Physical Review E* **1998**, *57*, 6825–6834.

[97]   W. Jahn and R. Strey, "Microstructure of Microemulsions by Freeze Fracture Electron Microscopy", *Journal of Physical Chemistry* **1988**, *92*, 2294–2301.

[98]   J. F. Bodet, J. R. Bellare, H. T. Davis, L. E. Scriven and W. G. Miller, "Fluid Microstructure Transition from Globular to Bicontinuous in Midrange Microemulsion", *Journal of Physical Chemistry* **1988**, *92*, 1898–1902.

[99]   M. S. Leaver, U. Olsson, H. Wennerstrom, R. Strey and U. Wurz, "Phase Behaviour and Structure in a Non-ionic Surfactant-Oil-Water Mixture", *Journal of the Chemical Society-Faraday Transactions* **1995**, *91*, 4269–4274.

[100]   U. Olsson and P. Schurtenberger, "Structure, Interactions, and Diffusion in a Ternary Nonionic Microemulsion Near Emulsification Failure", *Langmuir* **1993**, *9*, 3389–3394.

[101]   S. H. Chen, "Small-Angle Neutron-Scattering Studies of the Structure and Interaction in Micellar and Microemulsion Systems", *Annu. Rev. Phys. Chem.* **1986**, *37*, 351–399.

[102]   M. Kahlweit, R. Strey, R. Schomaecker and D. Haase, "General patterns of the phase behavior of mixtures of water, nonpolar solvents, amphiphiles, and electrolytes. 2", *Langmuir* **1989**, *5*, 305–315.

[103]   M. Teubner and R. Strey, "Origin of the scattering peak in microemulsions", *Journal of Chemical Physics* **1987**, *87*, 3195–3200.

[104]   I. S. Barnes, S. T. Hyde, B. W. Ninham, P. J. Derian, M. Drifford and T. N. Zemb, "Small-Angle X-Ray-Scattering from Ternary Microemulsions Determines Microstructure", *Journal of Physical Chemistry* **1988**, *92*, 2286–2293.

[105]   T. Tlusty, S. A. Safran, R. Menes and R. Strey, "Scaling laws for microemulsions governed by spontaneous curvature", *Physical Review Letters* **1997**, *78*, 2616–2619.

[106]   T. Sottmann and R. Strey, "Ultralow interfacial tensions in water-n-alkane-surfactant systems", *Journal of Chemical Physics* **1997**, *106*, 8606–8615.

[107]   M. Schwan, *Überkritische Mikroemulsionen zur Herstellung nanozellulärer Schäume – Principle of Supercritical Microemulsion Expansion (POSME)*, PhD-Thesis, University of Cologne, **2005**, Cologne.

[108]   E. S. J. Rudolph, M. J. Bovendeert, T. W. de Loos and J. de Arons, "Influence of Methane on the Phase Behavior of Oil + Water + Nonionic Surfactant Systems", *J. Phys. Chem. B* **1998**, *102*, 200–205.

[109]   R. von Hagen, *Mikroemulsionen mit überkritischen Fluiden für die technische Anwendung*, Diploma-Thesis, University of Cologne, **2009**, Cologne.

[110] K. Harrison, J. Goveas, K. P. Johnston and E. A. Orear, "Water-in-Carbon Dioxide Microemulsions with a Fluorocarbon-Hydrocarbon Hybrid Surfactant", *Langmuir* **1994**, *10*, 3536–3541.

[111] Y. Takebayashi, Y. Mashimo, D. Koike, S. Yoda, T. Furuya, M. Sagisaka, K. Otake, H. Sakai and M. Abe, "Fourier transform infrared spectroscopic study of water-in-supercritical CO_2 microemulsion as a function of water content", *Journal of Physical Chemistry B* **2008**, *112*, 8943-8949.

[112] M. Sagisaka, S. Iwama, S. Hasegawa, A. Yoshizawa, A. Mohamed, S. Cummings, S. E. Rogers, R. K. Heenan and J. Eastoe, "Super-Efficient Surfactant for Stabilizing Water-in-Carbon Dioxide Microemulsionst", *Langmuir* **2011**, *27*, 5772–5780.

[113] M. Sagisaka, D. Koike, S. Yoda, Y. Takebayashi, T. Furuya, A. Yoshizawa, H. Sakai, M. Abe and K. Otake, "Optimum tail length of fluorinated double-tail anionic surfactant for water/supercritical CO_2 microemulsion formation", *Langmuir* **2007**, *23*, 8784–8788.

[114] O. Holderer, M. Klostermann, M. Monkenbusch, R. Schweins, P. Lindner, R. Strey, D. Richter and T. Sottmann, "Soft fluctuating surfactant membranes in supercritical CO_2-microemulsions", *Phys. Chem. Chem. Phys.* **2011**, *13*, 3022–3025.

[115] P. W. Atkins, *Physikalische Chemie*, VCH Verlagsgesellschaft mbH, Weinheim 1990, 2. Auflage edn., **1990**.

[116] E. W. Lemmon, M. O. McLinden and D. G. Friend, "Thermophysical Properties of Fluid Systems", in *NIST Chemistry WebBook, NIST Standard Reference Database Number 69*, eds. P. J. Linstrom and W. G. Mallard, National Institute of Standards and Technology, Gaithersburg MD, 20899, 10.18434/T4D303, (retrieved January 5, 2018).

[117] A. Georgiadis, G. Maitland, J. P. M. Trusler and A. Bismarck, "Interfacial Tension Measurements of the (H2O + CO2) System at Elevated Pressures and Temperatures", *Journal of Chemical and Engineering Data* **2010**, *55*, 4168–4175.

[118] M. Sagisaka, T. Fujii, Y. Ozaki, S. Yoda, Y. Takebayashi, Y. Kondo, N. Yoshino, H. Sakai, M. Abe and K. Otake, "Interfacial properties of branch-tailed fluorinated surfactants yielding a water/supercritical CO_2 microemulsion", *Langmuir* **2004**, *20*, 2560–2566.

[119] M. Sagisaka, T. Fujii, D. Koike, S. Yoda, Y. Takebayashi, T. Furuya, A. Yoshizawa, H. Sakai, M. Abe and K. Otake, "Surfactant-mixing effects on the interfacial tension and the microemulsion formation in water/supercritical CO_2 system", *Langmuir* **2007**, *23*, 2369–2375.

[120] S. R. P. da Rocha and K. P. Johnston, "Interfacial thermodynamics of surfactants at the CO_2-water interface", *Langmuir* **2000**, *16*, 3690–3695.

[121] S. R. P. da Rocha, J. Dickson, D. Cho, P. J. Rossky and K. P. Johnston, "Stubby Surfactants for Stabilization of Water and CO2 Emulsions: Trisiloxanes", *Langmuir* **2003**, *19*, 3114–3120.

[122] J. Z. Zhu, L. Q. Chen, J. Shen and V. Tikare, "Coarsening kinetics from a variable-mobility Cahn-Hilliard equation: Application of a semi-implicit Fourier spectral method", *Physical Review E* **1999**, *60*, 3564–3572.

[123] O. Bayer, "Das Di-Isocyanat-Polyadditionsverfahren (Polyurethane)", *Angew. Chem.* **1947**, *59*, 257–272.

[124] D. Dieterich, "Polyurethane–nach 50 Jahren immer noch jung", *Chem. unserer Zeit* **1990**, *24*, 135–142.

[125] S. Geier, H. Schmitz, U. Göschel, P. Eyerer, A. Ostrowicki, N. Woicke, C. Ulrich, W. Lutz, J. Eschl, G. Rüb, M. Keuerleber, J. Diemert, J. Hauk, A. Stieneker, J. Woidasky, I. Fischer, K. Kretschmer, L. Ober, C. Kohlert, S. Ganslmeier, C. Schlade, H. Schüle, K. Kurz, K. U. Tönnes, P. Elsner, R. Protte, D. Liebing, A. Rodríguez, S. R. Raisch, H.-J. Dern, S. Schlünken, R. Bräuning and A. König, *Kunststoffe Eigenschaften und Anwendungen*, Springer-Verlag Heidelberg, Heidelberg Dordrecht London New York, **2012**.

[126] S.-T. Lee, C. B. Park and N. S. Ramesh, *Polymeric foams: science and technology*, CRC Press, **2006**.

[127] C. Six and F. Richter, Isocyanates, Organic, in *Ullmann's Encyclopedia of Industrial Chemistry*, Wiley-VCH Verlag GmbH & Co. KGaA, **2000**, 10.1002/14356007.a14_611.

[128] C. Mortimer, Thieme, Stuttgart, New York, 1987.

[129] United Nations Environment Programme Ozone Secretariat, *Handbook for the Montreal Protocol on Substances that Deplete the Ozone Layer*, Secretariat for The Vienna Convention for the Protection of the Ozone Layer & The Montreal Protocol on Substances that Deplete the Ozone Layer, Eleventh Edition (2017) edn., **2017**.

[130] R. A. Yourd, "Compression Creep and Long-Term Dimensional Stability in Appliance Rigid Foam", *Journal of Cellular Plastics* **1996**, *32*, 601–616.

[131] G. Rossmy, H. J. Kollmeier, W. Lidy, H. Schator and M. Wiemann, "Mechanism of the stabilization of flexible polyether polyurethane foams by silicone-based surfactants", *Journal of Cellular Plastics* **1981**, *17*, 319–327.

[132] W. Friederichs, personal communication.

[133] A. J. Siuta, W. E. Starner, B. A. Toseland and R. M. Machado, Google Patents, 1991.

[134] J. R. Quay and J. P. Casey, Google Patents, 1990.

[135] J. W. Gooch, Reaction Injection Molding, in *Encyclopedic Dictionary of Polymers*, ed. J. W. Gooch, Springer New York, New York, NY, **2011**, 611–611, 10.1007/978-1-4419-6247-8_9790.

[136] W. Kaiser, *Kunststoffchemie für Ingenieure: von der Synthese bis zur Anwendung*, Carl Hanser Verlag GmbH Co KG, **2015**.

[137] M. H. Pahl and E. Muschelknautz, "Einsatz und Auslegung statischer Mischer", *Chem. Ing. Tech.* **1979**, *51*, 347–364.

[138] M. H. Pahl and E. Muschelknautz, "Statische mischer und ihre anwendung", *Chem. Ing. Tech.* **1980**, *52*, 285–291.

[139] P. Lindner, ed., *Neutrons, X-rays and light : scattering methods applied to soft condensed matter*, Elsevier, Amsterdam [u.a.], 2002.

[140] P. N. Pusey, Introduction to scattering experiments, in *Neutrons, X-rays, and light: scattering methods applied to soft condensed matter*, eds. P. Lindner and T. Zemb, Elsevier, Amsterdam [u.a.], **2002**, ch. 1, 3-22.

[141] P. Lindner, Scattering experiments: experimental aspects, initial data reduction and absolute calibration, in *Neutrons, X-Rays and Light: Scattering methods applied to soft condensed matter*, eds. P. Lindner and T. Zemb, Elsevier, Amsterdam [u.a.], **2002**, ch. 2, 23-48.

[142] P. Schurtenberger, Contrast and contrast variation in neutron, X-ray, and light scattering, in *Neutrons, X-rays and Light: Scattering Methods Applied to Soft Condensed Matter*, eds. P. Lindner and T. Zemb, Elsevier, Amsterdam [u.a], **2002**, ch. 7, 127–144.

[143] P. Kienzle, Neutron activation and scattering calculator, https://www.ncnr.nist.gov/resources/activation/, Accessed 2018.

[144] I. Hoffmann, "Neutrons for the study of dynamics in soft matter systems", *Colloid Polym. Sci.* **2014**, *292*, 2053-2069.

[145] Institut Laue-Langevin, D11 – Lowest momentum transfer & lowest background small-angle neutron scattering instrument, https://www.ill.eu/instruments-support/instruments-groups/instruments/d11/description/instrument-layout/, Accessed 2018.

[146] Institut Laue-Langevin, The ILL High-Flux Reactor, https://www.ill.eu/reactor-environment-safety/high-flux-reactor/, Accessed 2018.

[147] Institut Laue-Langevin, LAMP, https://www.ill.eu/html/instruments-support/computing-for-science/cs-software/all-software/lamp/, Accessed 2018.

[148] Institut Laue-Langevin, Graphical Reduction and Analysis SANS Program for MatlabTM, https://www.ill.eu/instruments-support/instruments-groups/groups/lss/grasp/home/, Accessed 2018.

[149] Institut Laue-Langevin, D22 – a small-angle neutron scattering diffractometer, https://www.ill.eu/instruments-support/instruments-groups/instruments/d22/description/instrument-layout/, Accessed 2018.

[150] O. Glatter, R. Strey, K. V. Schubert and E. W. Kaler, "Small angle scattering applied to microemulsions", *Berichte Der Bunsen-Gesellschaft-Physical Chemistry Chemical Physics* **1996**, *100*, 323-335.

[151] K. V. Schubert and R. Strey, "Small.angle neutron scattering from microemulsions near the disorder line in water/formamide-octane-C_iE_j systems", *Journal of Chemical Physics* **1991**, *95*, 8532-8545.

[152] K. V. Schubert, R. Strey, S. R. Kline and E. W. Kaler, "Small-angle neutron-scattering near Lifshitz lines: Transition from weakly structured mixtures to microemulsions", *Journal of Chemical Physics* **1994**, *101*, 5343-5355.

[153] H. Leitao, M. M. T. da Gama and R. Strey, "Scaling of the interfacial tension of microemulsions: A Landau theory approach", *Journal of Chemical Physics* **1998**, *108*, 4189-4198.

[154] M. Kahlweit, R. Strey, M. Aratono, G. Busse, J. Jen and K. V. Schubert, "Tricritical points in water-oil-amphiphile mixtures", *J. Chem. Phys.* **1991**, *95*, 2842-2853.

[155] L. Golubovic, "Passages and droplets in lamellar fluid membrane phases", *Physical Review E* **1994**, *50*, R2419-R2422.

[156] P. Pieruschka and S. A. Safran, "Random Interface Model of Sponge Phases", *Europhysics Letters* **1995**, *31*, 207-212.

[157] P. Pieruschka, S. A. Safran and S. T. Marcelja, "Comment on 'Fluctuating interfaces in microemulsion and sponge phases'", *Physical Review E* **1995**, *52*, 1245-1247.

[158] G. Porod, General theory, in *Small angle X-ray scattering*, **1982**.

[159] R. Strey, J. Winkler and L. Magid, "Small-Angle Neutron-Scattering from Diffuse Interfaces .1. Mono- and Bilayers in the Water-Octane-$C_{12}E_5$ System", *Journal of Physical Chemistry* **1991**, *95*, 7502-7507.

[160] Y. Talmon and S. Prager, "Statistical thermodynamics of phase equilibria in microemulsions", *The Journal of Chemical Physics* **1978**, *69*, 2984-2991.

[161] P. Debye, H. R. Anderson and H. Brumberger, "Scattering by an Inhomogeneous Solid. II. The Correlation Function and Its Application", *J. Appl. Phys.* **1957**, *28*, 679-683.

[162] P. Debye and A. M. Bueche, "Scattering by an Inhomogeneous Solid", *J. Appl. Phys.* **1949**, *20*, 518-525.

[163] A. Guinier and G. Fournet, *Small-angle scattering of X-rays*, John Wiley and Sons, Inc., New York, **1955**.

[164] O. Glatter, The inverse scattering problem in small-angle scattering, in *Neutrons, X-Rays and Light: Scattering methods applied to soft condensed matter*, eds. P. Lindner and T. Zemb, Elsevier, Amsterdam [u.a.], **2002**, ch. 4, 73-102.

[165]   O. Spalla, General theorems in small-angle scattering, in *Neutrons, X-Rays and Light: Scattering Methods Applied to Soft Condensed Matter*, eds. P. Lindner and T. Zemb, Elsevier, Amsterdam [u.a.], **2002**, ch. 3, 49–71.

[166]   M. Kotlarchyk and S. H. Chen, "Analysis of small angle neutron scattering spectra from polydisperse interacting colloids", *Journal of Chemical Physics* **1983**, *79*, 2461–2469.

[167]   J. S. Pedersen, "Determination of Size Distributions from Small-Angle Scattering Data for Systems with Effective Hard-Sphere Interactions", *Journal of Applied Crystallography* **1994**, *27*, 595–608.

[168]   T. Foster, *Microemulsions as compartmentalised reaction media : structural characterisation of water in oil microemulsions*, PhD-Thesis, University of Cologne, **2007**, Göttingen.

[169]   M. Gradzielski, D. Langevin, T. Sottmann and R. Strey, "Droplet microemulsions at the emulsification boundary: The influence of the surfactant structure on the elastic constants of the amphiphillic film", *Journal of Chemical Physics* **1997**, *106*, 8232–8238.

[170]   M. Gradzielski, D. Langevin, L. Magid and R. Strey, "Small-Angle Neutron Scattering from Diffuse Interfaces. 2. Polydisperse Shells in Water-n-Alkane-$C_{12}E_4$ Microemulsions", *Journal of Physical Chemistry* **1995**, *99*, 13232–13238.

[171]   M. E. Cates and S. J. Candau, "Statics and dynamics of worm-like surfactant micelles", *Journal of Physics-Condensed Matter* **1990**, *2*, 6869–6892.

[172]   J. Bang, S. M. Jain, Z. B. Li, T. P. Lodge, J. S. Pedersen, E. Kesselman and Y. Talmon, "Sphere, cylinder, and vesicle nanoaggregates in poly (styrene-b-isoprene) diblock copolymer solutions", *Macromolecules* **2006**, *39*, 1199–1208.

[173]   T. Neugebauer, "Berechnung der Lichtzerstreuung von Fadenkettenlösungen", *Annalen der Physik* **1943**, *434*, 509–533.

[174]   T. Foster, "Universal Analytical Scattering Form Factor for Shell-, Core-Shell, or Homogeneous Particles with Continuously Variable Density Profile Shape", *Journal of Physical Chemistry B* **2011**, *115*, 10207–10217.

[175]   J. S. Pedersen, Modelling of small-angle scattering data from colloids and polymer systems, in *Neutrons, X-rays and Light*, eds. P Lindner and T. Zemb, Elsevier, Amsterdam [u.a.], **2002**, ch. 16, 391–420.

[176]   D. J. Kinning and E. L. Thomas, "Hard-Sphere Interactions between Spherical Domains in Diblock Copolymers", *Macromolecules* **1984**, *17*, 1712–1718.

[177]   J. K. Percus and G. J. Yevick, "Analysis of Classical Statistical Mechanics by Means of Collective Coordinates", *Physical Review* **1958**, *110*, 1–13.

[178]   M. Kotlarchyk, S. H. Chen and J. S. Huang, "Critical behavior of a microemulsion studied by small-angle neutron scattering", *Physical Review A* **1983**, *28*, 508–511.

[179] L. S. Ornstein and F. Zernike, "Acculental deviations of density and opalescence at the critical point of a simple substance", *Proceedings of the Koninklijke Akademie Van Wetenschappen Te Amsterdam* **1914**, *17*, 793-806.

[180] R. Zorn, Inelastic Neutron Scattering: Dynamics of Polymers, in *Neutrons, X-rays and Light: Scattering Methods Applied to Soft Condensed Matter*, eds. P. Lindner and T. Zemb, Elsevier, Amsterdam [u.a.], **2002**, ch. 10, 221–255.

[181] F. Mezei, ed., *Neutron spin echo spectroscopy : basics, trends and applications*, Springer, Berlin, 2003.

[182] Institut Laue-Langevin, IN15 – Spin-echo spectrometer with time-of-flight option, https://www.ill.eu/instruments-support/instruments-groups/instruments/in15/description/instrument-layout/, Accessed 2018.

[183] F. Mezei, The principles of neutron spin echo, in *Neutron Spin Echo: Proceedings of a Laue-Langevin Institut Workshop Grenoble, October 15–16, 1979*, ed. F. Mezei, Springer Berlin Heidelberg, Berlin, Heidelberg, **1980**, 1–26, 10.1007/3-540-10004-0_16.

[184] A. G. Zilman and R. Granek, "Undulations and dynamic structure factor of membranes", *Physical Review Letters* **1996**, *77*, 4788–4791.

[185] M. Mihailescu, M. Monkenbusch, H. Endo, J. Allgaier, G. Gompper, J. Stellbrink, D. Richter, B. Jakobs, T. Sottmann and B. Farago, "Dynamics of bicontinuous microemulsion phases with and without amphiphilic block-copolymers", *Journal of Chemical Physics* **2001**, *115*, 9563-9577.

[186] F. Brochard and J. F. Lennon, "Frequency Spectrum of Flicker Phenomenon in Erythrocytes", *Journal De Physique* **1975**, *36*, 1035–1047.

[187] R. Messager, P. Bassereau and G. Porte, "Dynamics of the undulation mode in swollen lamellar phases", *Journal De Physique* **1990**, *51*, 1329–1340.

[188] X. Trier, K. Granby and J. H. Christensen, "Polyfluorinated surfactants (PFS) in paper and board coatings for food packaging", *Environmental Science and Pollution Research* **2011**, *18*, 1108–1120.

[189] European Commission, Directorate-General for Internal Market, Industry and Entrepreneurship and SMEs, COMMISSION REGULATION (EU) 2017/1000 of 13 June 2017 amending Annex XVII to Regulation (EC) No 1907/2006 of the European Parliament and of the Council concerning the Registration, Evaluation, Authorisation and Restriction of Chemicals (REACH) as regards perfluorooctanoic acid (PFOA), its salts and PFOA-related substances 14/06/2017.

[190] Office of Pollution Prevention and Toxics U.S. Environmental Protection Agency, 2010/15 PFOA Stewardship Program Guidance on Reporting Emissions and Product Content, October 2006.

[191] ChemPoint, ZONYL® TO CAPSTONE® FLUOROSURFACTANT & REPELLENT TRANSITION GUIDE, https://go.chempoint.com/zonyl-capstone, Accessed 2018.

[192] M. Kahlweit, G. Busse and B. Faulhaber, "On the effect of acids and bases on water-amphiphile interactions", *Langmuir* **2000**, *16*, 1020–1024.

[193] M. Kahlweit, R. Strey and D. Haase, "Phase behavior of multicomponent systems water-oil-amphiphile-electrolyte. 3", *J. Phys. Chem.* **1985**, *89*, 163-171.

[194] H. Egger, T. Sottmann, R. Strey, C. Valero and A. Berkessel, "Nonionic microemulsions with chlorinated hydrocarbons for catalysis", *Tenside Surfactants Detergents* **2002**, *39*, 17–22.

[195] R. Schwering, D. Ghosh, R. Strey and T. Sottmann, "Sugar-Based Microemulsions as Templates for Nanostructured Materials: A Systematic Phase Behavior Study", *Journal of Chemical and Engineering Data* **2015**, *60*, 124–136.

[196] R. Wiebe and V. L. Gaddy, "The Solubility of Carbon Dioxide in Water at Various Temperatures from 12 to 40° and at Pressures to 500 Atmospheres. Critical Phenomena*", *Journal of the American Chemical Society* **1940**, *62*, 815–817.

[197] S. K. Ghosh, S. Komura, J. Matsuba, H. Seto, T. Takeda and M. Hikosaka, "Phase transition between microemulsion and lamellar phases in a C12E5/water/n-octane amphiphilic system", *Japanese Journal of Applied Physics Part 1–Regular Papers Short Notes & Review Papers* **1998**, *37*, 919–924.

[198] S. H. Chen, S. L. Chang and R. Strey, On the interpretation of scattering peaks from bicontinuous microemulsions, in *Trends in Colloid and Interface Science IV*, eds. M. Zulauf, P. Lindner and P. Terech, Steinkopff, **1990**, 30–35, 10.1007/BFb0115519.

[199] R. F. Tabor, J. Eastoe and I. Grillo, "Time-resolved small-angle neutron scattering as a lamellar phase evolves into a microemulsion", *Soft Matter* **2009**, *5*, 2125–2129.

[200] M. Nagao and H. Seto, "Small-angle neutron scattering study of a pressure-induced phase transition in a ternary microemulsion composed of AOT, D_2O, and *n*-decane", *Physical Review E* **1999**, *59*, 3169–3176.

[201] G. Gompper, H. Endo, M. Mihailescu, J. Allgaier, M. Monkenbusch, D. Richter, B. Jakobs, T. Sottmann and R. Strey, "Measuring bending rigidity and spatial renormalization in bicontinuous microemulsions", *Europhysics Letters* **2001**, *56*, 683-689.

[202] O. Holderer, H. Frielinghaus, M. Monkenbusch, M. Klostermann, T. Sottmann and D. Richter, "Experimental determination of bending rigidity and saddle splay modulus in bicontinuous microemulsions", *Soft Matter* **2013**, *9*, 2308–2313.

[203] R. Strey, *Zur Mikrostruktur von Mikroemulsionen*, habilitation thesis, University of Göttingen, **1992**, Göttingen.

[204] C. T. Lee, K. P. Johnston, H. J. Dai, H. D. Cochran, Y. B. Melnichenko and G. D. Wignall, "Droplet interactions in water-in-carbon dioxide microemulsions near the critical point: A small-angle neutron scattering study", *Journal of Physical Chemistry B* **2001**, *105*, 3540–3548.

[205] P. G. de Gennes, "Liquid dynamics and inelastic scattering of neutrons", *Physica* **1959**, *25*, 825–839.

[206] H. Egger, G. H. Findenegg, O. Holderer, R. Biehl, M. Monkenbusch and T. Hellweg, "Bending elastic properties of a block copolymer-rich lamellar phase doped by a surfactant: a neutron spin-echo study", *Soft Matter* **2014**, *10*, 6926–6930.

[207] S. Wellert, H.-J. Altmann, A. Richardt, A. Lapp, P. Falus, B. Farago and T. Hellweg, "Dynamics of the interfacial film in bicontinuous microemulsions based on a partly ionic surfactant mixture: A neutron spin-echo study", *The European Physical Journal E* **2010**, *33*, 243-250.

[208] O. Holderer, H. Frielinghaus, M. Monkenbusch, J. Allgaier, D. Richter and B. Farago, "Hydrodynamic effects in bicontinuous microemulsions measured by inelastic neutron scattering", *European Physical Journal E* **2007**, *22*, 157–161.

[209] L. R. Arriaga, I. Lopez-Montero, F. Monroy, G. Orts-Gil, B. Farago and T. Hellweg, "Stiffening Effect of Cholesterol on Disordered Lipid Phases: A Combined Neutron Spin Echo plus Dynamic Light Scattering Analysis of the Bending Elasticity of Large Unilamellar Vesicles", *Biophysical Journal* **2009**, *96*, 3629–3637.

[210] B. Farago, "Recent results from the ILL NSEs", *Physica B: Condensed Matter* **2007**, *397*, 91–94.

[211] O. Holderer, H. Frielinghaus, D. Byelov, M. Monkenbusch, J. Allgaier and D. Richter, "Dynamic properties of microemulsions modified with homopolymers and diblock copolymers: The determination of bending moduli and renormalization effects", *Journal of Chemical Physics* **2005**, *122*.

[212] N. Becker, *Polyurethan-Schäume – Neue Strategien zur Minimerung der Zellgröße*, PhD-Thesis, University of Cologne, **2014**, Cologne.

[213] S. R. P. da Rocha, P. A. Psathas, E. Klein and K. P. Johnston, "Concentrated CO_2-in-water emulsions with nonionic polymeric surfactants", *Journal of Colloid and Interface Science* **2001**, *239*, 241–253.

[214] T. A. Hoefling, R. M. Enick and E. J. Beckman, "Microemulsions in near-critical and supercritical carbon dioxide", *The Journal of Physical Chemistry* **1991**, *95*, 7127–7129.

[215] X. Li, R. M. Washenberger, L. E. Scriven, H. T. Davis and R. M. Hill, "Phase Behavior and Microstructure of Water/Trisiloxane E6 and E10 Polyoxyethylene Surfactant/Silicone Oil Systems", *Langmuir* **1999**, *15*, 2278–2289.

[216] G. Chemikalien and c. o. B. R. u. c. Industrie, Tris(1-chlor-2-propyl)phosphat, http://www.gischem.de/suche/dokument.htm?client_session_Dokument=2892.

[217] N. Becker, *Anti-Aging-Maßnahmen für Polyurethan-Nanoschäume*, Diploma-Thesis, University of Cologne, **2011**, Cologne.

[218] S.-Y. Tseng, *Formulation and Characterization of Polyol / Water-containing CO2 Microemulsions*, Master thesis (unpublished), University of Stuttgart, **2016**, Stuttgart.

[219] P. L. du Nouy, "An interfacial tensiometer for universal use", *J. Gen. Physiol.* **1925**, *7*, 625–631.

[220] J. W. Gooch, "Encyclopedic Dictionary of Polymers", **2007**.

[221] T. Svitova, R. M. Hill, Y. Smirnova, A. Stuermer and G. Yakubov, "Wetting and interfacial transitions in dilute solutions of trisiloxane surfactants", *Langmuir* **1998**, *14*, 5023-5031.

[222] J. Schuster and A. Bräuer, personal communication.

[223] H. Kuhn and S. Vogt, personal communication.

[224] L. Kramer, *Überkritische CO₂-Mikroemulsionen als Vorstufen für Nanoschäume – Darstellung, Charakterisierung und Nanostruktur*, PhD-Thesis, University of Cologne, **2008**, Cologne.

[225] SINTERFACE Technologies e.K., Ring/Plate Tensiometer STA, http://www.sinterface.com/products/measurement/tensiometry/ring_plate_tensiometer_sta /index.html, Accessed 2018.

[226] W. D. Harkins and H. F. Jordan, "A method for the determination of surface tension from the maximum pull on a ring", *J. Am. Chem. Soc.* **1930**, *52*, 1751–1772.

[227] Anton Paar GmbH, DMA™ 5000 M density meter, https://www.anton-paar.com/corp-en/products/details/dmatm-5000-m-density-meter/, Accessed 2018.

Curriculum Vitae

Name	Stefan Lülsdorf
Born	24.03.1990, Troisdorf, Germany
Marital status	single

Education

1996–2000	Elementary school: Evangelische Grundschule Viktoriastraße in Troisdorf, Germany
2000–2009	High school: Gymnasium Zum Altenforst in Troisdorf, Germany
	General qualification for university entrance (Grade 1.8)

Academic Education

10/2009–09/2012	Study of Chemistry, University of Cologne, Germany (Grade 1.7, Bachelor of Science)
07/2012–10/2012	Bachelor Thesis at the University of Cologne, Germany, on "*Hochviskose Zucker-Mikroemulsionen für Extrusionsprozesse*", advisor Prof. Dr. R. *Strey*
10/2012–12/2014	Study of Chemistry, University of Cologne, Germany (Grade 1.5, Master of Science)
05/2014–11/2014	Master Thesis at the University of Cologne, Germany, on "*Formulierung von neuen Polyol-Mikroemulsionen zur Synthese von Polyurethan-Nanoschäumen*", advisor Prof. Dr. R. *Strey*

Scientific Career

01/2015–06/2015	Research assistant and start PhD-thesis at the Institute of Physical Chemistry, University of Cologne, group of Prof. Dr. R. *Strey*
Since 06/2015	Research assistant and PhD-thesis at the Institute of Physical Chemistry, University of Stuttgart, group of Apl. Prof. Dr. T. *Sottmann*
10/2014–04/2018	Frequent research trips to the Institut Laue-Langevin, Grenoble, France

Scholarships

10/2011–09/2012	Deutschlandstipendium (founded by AMGEN GmbH)
10/2012–09/2013	Deutschlandstipendium (founded by AMGEN GmbH)